The Teamwork Advantage

The Teamwork Advantage

An Inside Look at Japanese Product and Technology Development

Jeffrey L. Funk

Foreword by Richard J. Schonberger

Publisher's Message by Norman Bodek

Productivity Press

CAMBRIDGE, MASSACHUSETTS
NORWALK, CONNECTICUT

Productivity Press
P.O. Box 3007
Cambridge, Massachusetts 02140
(617) 497-5146 (telephone)
(617) 868-3524 (telefax)

Cover design by Gary Ragaglia
Printed and bound by Maple-Vail Book Manufacturing Group
Printed in the United States of America on acid-free paper

Library of Congress Cataloging-in-Publication Data

Funk, Jeffrey L.
    The teamwork advantage : an inside look at Japanese product and technology development / Jeffrey L. Funk.
        p.   cm.
    Includes bibliographical references and index.
    ISBN 0-915299-69-0
    1. Work groups--Japan.   2. New products--Japan--Management.
3. Japan--Manufactures--Management.   I. Title.
HD66.F86   1992
658.4'036--dc20   92-10186
                                                   CIP

To Chris

Contents

Publisher's Message

What goes on when Japanese companies "do teamwork"? Much has been written about the remarkable recovery of Japanese industry in the years since World War II. The phenomenal success of companies such as Toyota and Matsushita has been well documented from the viewpoint of innovative manufacturing management systems such as just-in-time production. Beneath these systems, however, lies an essential foundation not always found in non-Japanese companies that attempt to adopt such methods. That foundation is a strong cultural basis for cooperative work to achieve a common purpose.

As important as teamwork has been on the production floor, it is perhaps even more essential in the product development and design organizations of Japanese companies. Teamwork dramatically reduces development lead times by standardizing the process and output and by bringing together people with appropriate expertise to work on particular aspects of a project. Teamwork also ensures that the customer's needs are incorporated into the design and that suppliers are in sync with the plan.

How does this teamwork actually happen? Western companies have long been aware that leading Japanese companies use a different approach in their development and design areas, but few details have been available about what they do on a daily basis and how such an approach might be adapted outside Japan. Jeff Funk rectifies the situation with *The Teamwork Advantage: An Inside Look at Japanese Product and Technology Development*. Dr. Funk, a member of the business administration faculty at Pennsylvania State University, is uniquely qualified to examine the day-to-day impact of Japanese corporate culture on teamwork, having spent 18 months working as an engineer within Japanese electronics companies.

Most of the firsthand observations in *The Teamwork Advantage* are based on Dr. Funk's year-long experience as a Westinghouse exchange engineer at Mitsubishi Electric's Fukuoka Works. Working alongside Japanese engineers in semiconductor production equipment development, he was able to observe the inner operations of the department and its interactions with internal suppliers and customers. In the process, he attended group meetings, went on assignments to other locations, and interviewed managers, engineers, and other employees about how they worked together. The result is the most comprehensive and valuable view yet available of the daily workings of a Japanese engineering organization.

Dr. Funk begins the book with a discussion of the reasons underlying Japanese manufacturing success. In the first two parts he examines previous explanations for this success, synthesizing them with his own observations about the role of teamwork. In Part Three, Dr. Funk identifies four primary elements of Japanese corporate culture that contribute to successful teamwork:

1. A product-oriented organization that emphasizes multifunctional employees, employee-developed procedures and plans, and shared responsibilities.
2. Visible management in the form of large public status boards, standard status forms, and shared public files.
3. Oral communication, in informal and formal communication networks, facilitated by the office layout.
4. Extensive training, including comprehensive training for newcomers and cross-functional job rotation.

The interactions of these powerful elements have a synergistic, reinforcing effect. The elements are referenced in specific examples throughout the next three parts of the book, describing work situations in key areas of the Mitsubishi semiconductor equipment department.

In Part Four, Dr. Funk gives some background on how teamwork helps Mitsubishi reduce cycle time and improve yield in the semiconductor factory; he also describes the interface between the internal equipment producers and users in developing computer integrated manufacturing systems. Although production process improvements have been described in other publications, Funk uses this part of the book to give a unique look at the roles of engineers, operators, and others in improving semiconductor production.

Part Five of the book offers an inside view of design office activities in a large Japanese electronics company, focusing on the coordination of the various groups involved in designing the software and hardware components of semiconductor equipment. Standard design methodologies are critical for rapid and uncomplicated meshing of these components; Dr. Funk describes mechanical and electrical design methods and devotes an entire chapter to software standardization methods in the semiconductor equipment department at Mitsubishi. Since a large percentage of equipment manufacturing cost can be traced to design, he also describes how design engineers work closely with manufacturing engineers, suppliers, and technology lab employees to keep the equipment inexpensive.

In Part Six, the setting moves from the office to the Mitsubishi laboratories and factories to describe the process of developing new technology for semiconductor equipment. The initial chapter presents a detailed look at how Mitsubishi determines which R&D projects to fund. This corporate-level decision process involves complex strategic considerations that shape the company's preparedness to enter growing new markets. Other chapters deal with how Mitsubishi's semiconductor equipment business determines the technological needs of its customers, how the laboratories improve technology incrementally as it is developed, and how a business develops, installs, and improves equipment in partnership with its internal customer. Dr. Funk also sheds light on the significance of horizontal and vertical integration, a feature of many large Japanese manufacturers, and underscores the importance of long-term relationships.

Throughout the book, Dr. Funk emphasizes how job rotation produces a network of managers who know where in the organization to look for the expertise they need for a project; much of their success is due to their ability to link up their own engineers with the outside specialists who can help expedite the development and design work, sometimes sending people to other locations for short or longer periods. The Mitsubishi enterprise appears to be organized to accommodate and encourage such job rotation and working recombinations of personnel.

Dr. Funk closes the book with a look at the downside of Japanese social and corporate culture — some of the reasons a Western company would not want to adopt the Japanese approach wholesale. There are, however, many aspects we could glean to improve teamwork in our organizations; Dr. Funk summarizes these in the final chapter in eight prescriptions for "becoming better at teamwork."

As Richard Schonberger points out in his foreword, *The Teamwork Advantage* makes a significant contribution to the management of product and technology development. Whether you are involved in product design, technology development, or manufacturing production, Dr. Funk's firsthand perspective will give you new insight on coordinated approaches to the develop-design-build cycle in your own organization.

We would like to express our gratitude to Jeff Funk for allowing us to publish this book. Dr. Schonberger's early review of the manuscript and contribution of a foreword are both very much appreciated.

Many people at Productivity worked together to produce this book. Thanks to Productivity, Inc. development manager Connie Dyer for early appraisal of the manuscript; to Productivity Press series editor Diane Asay for editorial review; to managing editor Dorothy Lohmann for supervising manuscript preparation, Laura St. Clair for word processing and proofreading, Dana Wilson for copyediting, and Jennifer Cross for indexing; to production manager David Lennon for supervising composition and proofing, Karla Tolbert for art preparation, and Caroline Kutil for typesetting; to Gary Ragaglia for cover design.

Norman Bodek
President, Productivity, Inc.

Karen R. Jones
Editor, Productivity Press

Foreword

Teamwork in business has little or nothing to do with national culture. It is just good management and discipline. That teamwork is better developed in Japanese industry is mainly because top-notch Japanese companies have been at it longer.

While teamwork is admirable in and of itself, its greater value may be in the innovative product and process ideas that it spawns. Jeffrey Funk shows that to be the case repeatedly throughout *The Teamwork Advantage*, a superior contribution in the management of product and technology development.

The main setting for Dr. Funk's narrative is a product development unit of Mitsubishi Electric Corporation in Japan — an excellent choice, since Mitsubishi has long been renowned for its superior development of teams of development engineers (I commented on its approach in 1982 in *Japanese Manufacturing Techniques: Nine Hidden Lessons in Simplicity*). After a brief scenario to set the stage for the book, the reader is introduced to the principles that underlie teamwork in Japanese companies, then taken through a procession of insights: how to train; how to move skilled people around and bring them together with complementarily skilled people; how to capture, file, and reuse information (including computer software modules) that in Western companies is often jealously guarded in the head or file drawer of the individual specialist; how to control (and measure) without inhibiting initiative; and how to bring projects to successful, early completion.

These are examples of *how to make teams work,* and we see that Mitsubishi's practices have been improved, standardized, codified, taught, and then improved some more, year after year. Some practices are, of course, borrowed from other firms going through a similar trial-and-error evolution. As Dr. Funk confirms, numerous other superior Japanese companies have been following a similar path in perfecting product development and team management processes.

This is not to say that Western companies are blindly sticking with the separation of experts into functional specialties. Multifunctional design teams are common — and fast becoming the rule — in better managed Western manufacturing companies (and increasingly in service companies). In fact, Shin-ichi Nakagawa, a Mitsubishi engineer and leader of a group of 250 engineers from Japanese supplier companies working on design teams for Boeing's new 777 passenger airplane, is impressed with Boeing's approach. Nakagawa says that Japanese companies "are familiar with teams of design and production engineers, but they haven't experienced Boeing's all-embracing teams that include customers, suppliers, and support teams (Jeremy Main, "Betting on the 21st Century Jet, "*Fortune,* April 20, 1992, pp. 102-17).

No doubt, Nakagawa will take this (or any) lesson learned in his Boeing tour back to Mitsubishi in Japan, and its best features will be absorbed. Therein lies a serious, continuing weakness of Western companies: the lack of disciplined mechanisms for adapting and adopting — and avoiding reinvention of the wheel. Jeffrey Funk makes this plain, and his plain talk — with convincing examples — may have the necessary shock value to set the corrective wheels in motion.

Richard J. Schonberger
Schonberger & Associates, Inc.

Preface and Acknowledgments

My first exposure to Japan was Ezra Vogel's book, *Japan as Number One*, in 1980. Although I found the book very interesting, like most Americans at that time I still saw Japan not only as completely peripheral to my life, but also as a distant second to the United States economically. Several years and many books on Japan later, however, I was given two opportunities to live in Japan that drastically changed my perspectives on these matters.

In the summer of 1984, while completing a doctoral dissertation at Carnegie-Mellon University on the economics and management of robotic assembly, I learned that the American Electronics Association was offering U.S. engineers an opportunity to work for a Japanese electronics company in Japan. Since by this time Japan was the undisputed leader in the application of robotics, I quickly applied for the program and was fortunate to be one of seven engineers chosen. The American Electronics Association put us through a two-month intensive language program and I was given a six-month position at Yokogawa

Hokushin Electric (since split into Yokogawa Electric and Hokushin Precision) in its corporate marketing department.

Even on my first day at Yokogawa, it was obvious that Japanese firms were different from U.S. firms (I had worked for Hughes Aircraft before entering graduate school and had visited many U.S. factories and offices as part of my research in graduate school). I certainly couldn't say at that point that Japanese firms were better than U.S. firms, but it was obvious that they were different. In fact — as I attempt to describe in Part One through the eyes of a first-time visitor to Japan — I was initially struck more by the disadvantages than the advantages of the Japanese office. Japanese offices are crowded and loud, few people have personal computers or personal telephones, and the facilities are spartan, to say the least.

My experience at Yokogawa gave me the opportunity to see a number of Japanese factories and offices. Yokogawa, the second-largest supplier of process and factory control systems in the world, wanted to sell its factory control systems in the United States. To improve its understanding of the U.S. market for its products, its management asked me to compare Japanese and U.S. approaches to factory automation. As part of this assignment, I visited more than 30 Japanese factories, including those of Toyota, Asahi, and Isuzu, and interviewed (primarily in English) numerous manufacturing managers and engineers about their approach to factory automation. Through meetings and other activities in the corporate marketing department, I also learned a great deal about how a Japanese office works.

After returning to the United States, I knew that I had learned some important things but didn't know how to describe their significance to my new colleagues at the Westinghouse R&D Center, where I worked on the development of tools to support design for manufacturing. Fascinated by Japanese manufacturing techniques, culture, history, and management, I read every book I could find on these subjects. I also continued to study the Japanese language in preparation for another, albeit still unplanned, visit to Japan.

An opportunity presented itself in the summer of 1988 when I was selected for the Westinghouse-Mitsubishi engineer exchange program. This program included a one-year engineering position in a Mitsubishi division and two months of intensive language study at the University of Pittsburgh. Although the language program didn't provide enough vocabulary to make me fluent, it covered all of the grammar necessary to speak and read Japanese. Before I left for Japan, I decided to speak only Japanese in my work at Mitsubishi. Due to the patience of many Mitsubishi employees, I was able to meet this objective except when I met a fluent English speaker (this was actually quite rare where I was located) or when I returned the favor of Japanese lessons with English lessons.

I worked at Mitsubishi's Fukuoka Works, which is located on Kyushu island. Japanese use the term "works" (*seisakusho*) to describe a factory that also has design engineering and other functions, whereas a "factory" (*kojo*) performs only manufacturing operations. The Fukuoka Works originally produced only motors, hoists, and other industrial equipment. Due to the oil price increases in the 1970s, however, the growth rate for these markets began to slow considerably and the works was faced with overcapacity and reduced profits. To avoid laying off employees, Mitsubishi's corporate headquarters decided to begin producing semiconductor chips at the Fukuoka Works using some of those employees. Production workers and manufacturing engineers at the Fukuoka Works were given substantial training in semiconductor production and were transferred to this new part of Mitsubishi's semiconductor business in 1981. In 1989, only 15 percent of the factory's sales output were in the original markets served by the works.

In 1985, Mitsubishi formed the semiconductor equipment department (SED) at the Fukuoka Works; this is the department where I worked in 1989. This department produces much of the equipment now purchased by Mitsubishi's semiconductor factories. The SED works closely with several other parts of Mitsubishi's semiconductor business to develop and introduce new equipment into the company's semiconductor factories; these other parts include the headquarters for Mitsubishi's semiconductor business, the semiconductor factories, and several of the laboratories. Therefore, in addition to analyzing how the semiconductor factories and the SED operate, I also had the opportunity to analyze how they work together and with these other elements of the Mitsubishi semiconductor business.

I worked in the wire bonder equipment group, one of seven equipment groups and ten total groups in the department. My assignment was to develop software for the wire bonder's vision system. This assignment gave me the opportunity to work with Japanese engineers on a daily basis. I attended and participated in most of the weekly wire bonder equipment meetings, the group's daily standup meetings, many informal meetings, and meetings with laboratory personnel — over 200 meetings in all. Due to the open office, I was able to see how Japanese employees spend their time and whom they talk to. Due to my semi-fluency and the open office, I was also able to overhear numerous conversations each day — either one side of a telephone call or conversations between employees who sat near me. I saw almost every document that was passed out at a meeting or distributed among employees. I spent many hours translating these documents (more than 100 pages) and asking questions about them.

Although the documents and meetings precipitated many of my interviews, I had a very good idea of what I wanted to learn about before my first day at Mitsubishi. Through books, journal, and magazine articles and my previous experience at Yokogawa, I had formed several hypotheses about Japan and Japanese management. I was convinced that the greater use of such strategies as JIT manufacturing, total quality control, focused factories, concurrent engineering, short product development cycles, and close relationships with suppliers, customers, and laboratories was an important reason for Japan's economic success. However, I didn't know which of these techniques were more important (I still don't know), and I didn't know how and why Japanese firms are able to implement these strategies better than U.S. firms.

I spent the majority of my time trying to understand how Mitsubishi and the SED implement these strategies. I conducted more than 200 interviews with employees from the SED, suppliers, the laboratories, and the factory, and translated more than 100 pages of documents. Appendix A describes these sources of data in more detail. In my two extended stays in Japan, I visited more than 50 Japanese factories and more than 10 Japanese laboratories. In addition, I have supplemented my data with interviews of more than 30 U.S. engineers who have worked for extended time periods in several Japanese manufacturing firms.

Midway through my experience at Mitsubishi, I concluded that teamwork was an important reason for Mitsubishi's success in implementing JIT manufacturing, total quality control, focused factories, concurrent engineering, short product development cycles, and close relationships with suppliers, customers, and laboratories. The successful implementation of each of these strategies requires a great deal of cooperation; it became very clear that the employees of the SED worked closely together and with their customers, suppliers, and laboratories on all of these strategies. It also became clear that Mitsubishi was doing a number of things differently from Westinghouse,

Hughes, and other U.S. firms (based on what is described in the management literature) and that these things were facilitating teamwork. (I also found that Yokogawa was very similar to Mitsubishi in those things that facilitate teamwork.) I refer to a framework of four basic elements of Japan's corporate culture to describe the teamwork-facilitating activities that distinguish Japanese firms from U.S. firms.

I realize that I have made generalizations about Japanese firms based on a limited scope of data. Since most of my data are from one firm (really one business within that firm), my generalizations are primarily based on data collected by other researchers. I also lack data from a comparable U.S. business. Although I worked two years as a semiconductor process engineer for Hughes Aircraft and five years as a researcher at Westinghouse, Hughes primarily serves the U.S. defense industry, and the Westinghouse R&D Center is not comparable to the SED. Therefore most of my comparisons are based on books and articles written about U.S. management.

I realize that this approach is problematic. Without multifirm comparisons between Japanese and U.S. companies, it is dangerous to draw conclusions about the exact reasons for Japan's manufacturing success vis-à-vis the United States. My intention, however, is neither to describe how all Japanese firms work nor to provide the exact reasons for their economic success. Rather, the purpose of this book is to guide the thoughts of practitioners and the studies of researchers. I believe that the elements of Japan's corporate culture presented in this book help Japanese firms and can help Western firms implement JIT manufacturing, total quality control, focused factories, concurrent engineering, short product development cycles, and close relationships with suppliers, customers, and laboratories. I hope that this book will help practitioners understand how to implement these strategies and help researchers identify hypotheses that should be tested in future studies. Further studies will be needed to determine the relative importance of each element of Japan's corporate culture (i.e., their percent contribution) to the implementation of these strategies.

As with most books, I am indebted to a large number of people, both American and Japanese, particularly in Yokogawa Hokushin Electric and Mitsubishi Electric Corporation. I am grateful to the American Electronics Association and to Westinghouse for the opportunities to live and work in Japan. At Yokogawa Hokushin Electric, I am particularly thankful to Nagakazu Shimizu for his time and patience. At Mitsubishi, I am indebted to many employees of the semiconductor equipment department, all of the members of the wire bonding equipment group, and several employees of the laboratories and corporate engineering. In particular, I would like to thank Messrs. Hirayama, Honda, Kazuhiro Kawabata, Koji Matsuda, Ebina and Matsumoto, Kishida, Tosh Yanagisawa, Banjo, Ishizuka, Hiroki, and Yokoyama.

I am grateful to Richard Schonberger, who took the time to read my manuscript and offer his comments. I would also like to thank Karen Jones, my editor at Productivity, for the many improvements she made to the manuscript. Any mistakes that may be found in the book are strictly my responsibility.

I am particularly indebted to my wife, Christina Haas, who, in addition to having a great deal of patience while I wrote the book, put her academic career on hold for almost two years so that we could live in Japan.

PART ONE

Explaining Japan's Manufacturing Success

Jay Edwards, a professional engineer from the United States, was visiting a Japanese electronics company for an intensive exchange study.* The program in which he was participating would give him an opportunity to observe firsthand the inner workings of the engineering offices of the company, a world leader in its technology.

The first thing Jay noticed when he arrived for a tour of the area where he would be working for the next month was the office itself. It was a large, open room that held about 100 desks, pushed up against one another in rather tight rows. The engineers — all of whom were men — had very little work space on top of the desks, little storage space within them, and no partitions for privacy. The walls were covered with slogans, charts, graphs — and bookcases filled with looseleaf binders. There were no computers in sight. "This isn't high technology!" Jay thought.

The office was also buzzing with activity — people moving here and there, conversations stopping and starting. It was loud — meetings seemed to be going on everywhere. Several phones seemed to be ringing simultaneously, although no one had a personal phone. A nearby engineer answered one telephone and then started yelling for another engineer who was wanted on the incoming call. Although they found the person right away, Jay found the yelling very distracting. "How can they get any work done?" he thought. "I wouldn't be able to think in a place like this."

Jay told his guide, Mr. Hirayama, that during his one-month stay he would like to learn about the department's design methodology and its approach to a technique called concurrent engineering. Hirayama smiled and excused himself and shortly

---

* Jay Edwards is a composite character based on my own observations and reactions and those reported by other American visitors to Japan.

returned with one of the many looseleaf binders from the shelves that lined the office walls. Hirayama began to show Jay flowcharts and detailed descriptions of the documents to be developed during each step in the design cycle. Jay was surprised; he really hadn't expected such a quick response — or such a detailed one. He was more surprised, however, at the detailed procedures that Hirayama was explaining and by the fact that they were developed and followed by the designers. "All these procedures!" Jay thought. "They seem awfully bureaucratic." Jay doubted if the engineers in his U.S. firm would follow such procedures — and if they did, he wondered, wouldn't they get in the way of creative solutions to problems?

Suddenly a bell rang and Hirayama excused himself again, as engineers in the office began cleaning up. They weren't just straightening up their desks: Mops, brooms, and pails were brought out of their storage places. Everyone seemed to be involved. As someone nearby emptied a trash basket into a plastic bag, Jay noticed that the basket was a discarded oil can. "One of the richest countries in the world, and they recycle their oil cans as office wastebaskets," he mused.

As Jay continued to watch the engineers-turned-janitors, he noticed that several of the managers were leaving. Jay asked Hirayama if these managers had late meetings that they were attending. "Maybe," Hirayama replied, "but they are more likely going home or out with friends." Jay couldn't hide his surprise. "But I thought you all worked so much overtime!" Hirayama explained that while he and most of the other engineers worked overtime, few of their managers did. "So," Jay thought, "the managers go home early and the regular employees work overtime. That's surprising too." Hirayama confided with a shy smile that he and the other engineers felt that it was easier to get work done without the managers around. Jay wondered how things could get done without managers. As he looked around, however, he realized that it was 30 minutes past quitting time and although there were no managers around, almost all of the engineers were still in the office working.

Although Jay himself planned to leave the office soon, he wanted to find out something about his new friend. When Jay asked him about his job, Hirayama said, "I'm a freshman." He went on to explain that new employees just out of college are called freshmen; they spend the first year of employment learning about the company and their department and receiving technical training, some of it one-on-one with senior engineers. Again Jay was surprised. "Don't you have a college degree?" he asked. "Oh yes," Hirayama replied, looking a little embarrassed. "But we really don't learn much that is useful in our jobs while we're in college. After working so hard in high school to get into a good university, we have to take it easy!"

Welcome to Japan. Like many Westerners who have visited Japanese companies, Jay found a number of things he didn't expect. "It seems like a developing country, instead of our chief competition," Jay told his wife on the phone that night. Indeed, Japanese firms seem to violate many American concepts of "good management." Although Jay Edwards is a hypothetical character, many Americans (including myself) are extremely surprised on their first visit to a Japanese firm because of several unusual characteristics: loud, crowded offices; few computers; no desk space; excessive procedures; no janitorial staff; managers who don't seem to work as hard as the regular employees or to be on top of what's going on; a Japanese university education that helps young people find jobs, but doesn't seem to prepare them for the work they will do.

However, as Americans know all too well, Japan is no longer a developing country. The Meiji Restoration, which occurred in 1867, was the beginning of Japan's long road from a third-world existence to its current status as one of the richest countries in the world. The Meiji Restoration instituted a better educational system, opened the country to Western ideas, and channeled funds toward industrial development. Although World War II slowed down Japan's economic growth quite a bit, in many ways the economic performance in those years has become but a small blip in the economic growth of Japan since the Meiji Restoration. It's important to keep in mind how far Japan's economy has progressed over the last 120 years when we discuss the sources of Japan's economic growth.

CHAPTER ONE

A Critical Look at Japan's Economic Success

S INCE World War II, Japan's economy has done well by almost every measure of economic performance. Its productivity increased almost 10 percent per year until the mid-1970s. Since then, its percentage increases have slowed down to only about 5 percent a year. It has a trade surplus in almost every manufacturing industry, despite having a per capita income that is now higher than that in the United States.

Americans used to compare themselves to the Japanese using unadjusted per capita income. It made us feel better that although our steel, auto, and other industries couldn't compete with Japan, we had higher wages. We told ourselves that Japan was doing well because its workers received lower wages. Since Japan's per capita income has surpassed that of the United States, however, we've started focusing on the "purchasing power" of that income. Japan's high prices, particularly on imported products, food, and housing, cause the Japanese to have a lower purchasing power than Americans and thus to have a lower standard of living.

However, it's a mistake to look at the static case. Japan may have a lower standard of living now, but that measure is increasing faster than the same standard in the United States. Japan's higher growth in productivity and per capita income will cause its standard of living to eventually surpass that of the United States. Propelling this productivity growth is an economy that is evolving from the production and export of low-technology products to higher technology products. Since higher technology industries tend to pay higher wages than lower technology industries, the higher technology industries have a higher level of productivity when measured in terms of dollars per labor-hour. Wages in the U.S. computer industry, for example, are about 25 percent higher than average wages in the manufacturing sector and almost 60 percent higher than the average wages in the U.S. economy as a whole.[1]

Japan's Increasing Strength in High-Technology Industries

Japan's economy has been evolving from the production of low-technology products to high-technology products for at least the last 20 years — and probably since the Meiji Restoration. Table 1-1 shows the contribution to Japan's exports by a number of different high-technology industries since 1967. The industries are organized according to the each industry's technology "intensiveness." The ranking was developed by the National Science Foundation, and it is based upon the percentage of sales that the industry spends on research and development.[2]

As shown in Table 1-1, high-technology products represented a larger percentage of Japan's exports in 1987 than in 1967. Their contribution to Japan's exports has increased from 12.4 percent in 1967, to 16.0 percent in 1974, to 17.2 percent in 1982, and to 29.6 percent in 1987. During the same time period, Japan's total exports increased from $10.4 billion in 1967 to $229.2 billion in 1987. The evolution of Japan's economy from low-technology

Table 1-1. Percentage of Japan's Exports by High-Technology Industry

Industry	High Technology			
	1967	1974	1982	1987
Aircraft[1]	<.1%	<.1%	<.1%	.1%
Office machines[2]	.7	1.3	3.3	7.9
Electrical components[3]	.5	.7	.58	4.6
Telecommunications equip.[4]	1.3	.9	3.2	5.5
Other electrical equip.[5]	.2	1.8	1.9	2.1
Pharmaceuticals[6]	.3	.2	.2	.2
Plastics[7]	2.8	3.1	1.3	1.5
Instruments[8]	<.1	.6	1.3	1.7
Photographic & optical equip.[9]	3.0	3.1	3.3	3.4
Indus. & agricultural chem.[10]	3.6	4.3	2.1	2.2
Total	12.4%	16.0%	19.0%	29.6%

1. SITC 792   2. SITC 75   3. SITC 776   4. SITC 772, 773   5. SITC 764
6. SITC 54   7. SITC 87   8. SITC 87   9. SITC 88   10. SITC 51, 52, 56, 591

Note: SITC = Standard International Trade Classification

to high-technology products is even reflected in the high-technology industries shown in Table 1-1. The percentage contribution to Japan's exports has dropped in the "bottom half" of the high-technology industries shown in Table 1-1 from about 9.7 percent in 1967 to 8.7 percent in 1987. The percentage contribution to Japan's exports from the "top half" of these industries has increased from 3.4 percent in 1967 to 20.1 percent in 1987. This is exactly what you would expect from a country that has been basically "catching up" to the rest of the world.

Table 1-1 says more about the Japanese economy than a look at any one year. When you look only at Japan's trade data for a given year, it is natural to assume that Japan does well only in those industries in which it has a large trade surplus. This perspective, however, has prevented Americans from noticing the continual evolution of Japan's economy from lower technology products to higher technology products. In the 1960s, the United States didn't mind if the Japanese dominated the steel, shipbuilding, and textile industries — we knew they couldn't make higher technology products such as automobiles and consumer electronics as well as we could. By the mid-1970s, however, Americans were enthusiastically buying Japanese-made cars, televisions, and other consumer electronics equipment. We still felt that the Japanese couldn't make high-technology products that required innovation and creativity, like semiconductors, instruments, and computers. By the mid-1980s, though, Japan controlled the world market for memory chips.

You would think that we would have learned by then. However, we continued to assume that Japanese firms would be unsuccessful in higher technology industries. We thought Japan couldn't make low-volume chips — but in 1986, three of the world's five largest suppliers of custom-designed chips were Japanese firms.[3] Now some people now claim that Japanese companies won't succeed in the computer industry because they don't yet control the market for personal computers or that they can't write software because they don't lead the world in it now. The conclusions about personal computers and software may be correct, but the reasons for them are wrong. If history is any guide, the conclusions may also be wrong.

Other misconceptions arise from looking at the trade data for only a given year. Since Japanese firms first developed a strong export capability in high-volume standardized industries such as steel, automobiles, and consumer electronics, many Americans believe that Japanese firms do well only in these types of industries. However, Japan's initial strength in these industries is probably due to their low technological requirements rather than to a lack of ability to compete in low-volume industries. Since low-volume industries typically produce customized products, they typically require higher spending on research and development.

As Table 1-1 shows, Japanese firms are becoming stronger in many low-volume customized industries that require high spending on research and development. Instruments, office machines, telecommunications equipment, and electrical equipment have much smaller product volumes and require much higher levels of customization than steel, automobiles, and consumer electronics. Other evidence for the strength of Japanese firms in lower volume industries also exists. Japanese automobile firms produce an average of only 500,000 vehicles per model, while U.S. firms produce 2.1 million vehicles per model.[4] Japanese firms have become the world's best producers of production machinery, a very low-volume product. Michael Porter found that Japan has a world-class export capability in 22 machinery industries versus 18 for the United States. Japan gained export share in 29 machinery industries between 1978 and 1985 and lost export share in only two machinery industries during this period.[5] Other studies have also concluded that Japanese firms are successful in producing nonstandardized products.[6]

Japan's Strength in the Discrete Parts Industries

Table 1-1 tells us several more interesting things. First, it says that there are a few industries in which Japan is not doing as well. Since chemicals, plastics, and pharmaceuticals are less technology-intensive than electrical equipment, telecommunications, semiconductors, and computers, it would be expected that Japan's competitive strength should be increasing in these industries, not decreasing. As Japan's exports were evolving from low-technology to high-technology products, it seems reasonable that the contribution from the chemical, plastic, and pharmaceutical industries would also have been increasing.

However, their percentage contribution to Japanese exports has been decreasing over the last 20 years; in fact, Japan had a slight trade deficit in the industrial chemical, agricultural chemical, and pharmaceutical industries in 1987.

The chemical, plastic, and pharmaceutical industries have a great deal in common with other lower technology industries in which Japan does not have the competitive advantage. Porter argues that Japanese firms have not been successful in the food and beverage, packaged consumer goods, textiles, or apparel industries. These industries are similar to the chemical, plastics, and pharmaceutical industries in that none of them makes products containing discrete parts. They are sometimes called the "process industries." In this book, I will refer to process industries as *nondiscrete parts industries*. Potential reasons for Japan's weaknesses in these industries will be described when I discuss the apparent strength of U.S. firms in many of these industries.

A second interesting observation about Table 1-1 is that Japan's exports are coming more and more from several high-technology industries with products such as instruments, telecommunications, electrical components (including semiconductors), and office equipment (including computers). All of these industries are part of the discrete parts industries. As a percentage of Japan's exports, office equipment more than doubled between 1982 and 1987. The percentage contribution from electrical components increased by a factor of five between 1982 and 1987. While the percentage contribution from office equipment and electrical components was increasing, Japan's total exports were also increasing by more than 50 percent between 1982 and 1987.

There are many more signals of Japan's increasing strength in the high-technology industries that are classified as discrete parts industries. In *Created in Japan,* Sheridan Tatsuno describes how Japanese firms have developed and continue to develop new technology in virtually every field of science and engineering related to electronics.[7] Not only does Japan receive 18.6 percent of the patents granted in the United States (up from 8.8 percent in 1975), but in terms of patent creativity, seven of the top ten computer and electronics firms are now Japanese.[8] Comparisons between specific American and Japanese technologies also show Japan ahead, or moving ahead, in almost every technology related to electronics and computers.[9] Further, the Japanese are even developing strength in the aircraft industry. According to the chairman of Boeing, Japan is working hard on the supersonic transport technology required for the next generation of commercial airplanes.[10] Clyde Prestowitz claims that 15 percent of the 767's parts came from Japan, as will 25 percent of the parts for the 7J7, the next generation of aircraft.[11]

Loss of Competitiveness in U.S.
High-Technology Industries

The U.S. economy is in many respects a mirror of Japan's. Firms in the United States are having problems in those industries in which Japanese firms have been succeeding. The U.S. trade deficit has migrated from low-technology industries to high-technology industries over the last 20 years. The deficit used to be only in steel, apparel, televisions, radios, and automobiles. In the 1980s it was in semiconductors, telecommunications equipment, and electrical machinery. Perhaps in the 1990s it will be in office equipment and computers.

I use a different type of ratio to evaluate U.S. strength in high-technology industries than I used to evaluate Japan's strength in these industries. For Japan, the percentage contribution of an industry's exports to Japan's total exports shows Japan's strength in specific high-technology industries (see Table 1-1). This measure would be somewhat misleading for the United States, however, since it has experienced a declining trade balance in virtually every manufacturing industry. This measure wouldn't tell us how U.S. high-technology industries are performing vis-à-vis the rest of the world. It would merely tell us that high-technology industries are doing better than lower technology industries in the United States.

A ratio of net trade (exports − imports) to total trade (exports + imports) more accurately shows the strength of U.S. high-technology industries. Called an import-export ratio, this index combines a measure of export strength with import penetration.[12] An industry with zero imports will have an import-export ratio of 1 and an industry with zero exports will have an import-export ratio of −1. An industry with balanced trade (exports equal to imports) will have a ratio of 0.

This ratio is a better measure of performance for a specific industry than the percentage contribution of that industry to the country's total exports. Since Japan has never imported many manufactured goods, however, its import-export ratios have always been close to 1 for almost every industry. Therefore, the percentage contribution to exports was used to evaluate Japan's strength in several high-technology industries. The United States, on the other hand, has a relatively open market in which the level of imports and exports largely reflects the capability of U.S. and foreign firms.[13]

Figure 1-1 shows the ratio of net trade to total trade for U.S. high-technology industries. This figure shows that U.S. firms have lost ground in almost every high-technology industry over the last 20 years. Many of these increased imports and decreased exports have resulted from competition from Japanese firms. The United States has been increasing its imports of high-technology products from Japan, particularly in the industries producing photographic and optical products, instruments, semiconductors, telecommunications equipment, electrical components, and office machines. Even computers, one segment of office machines, posted only a $2 billion surplus in 1987 on exports of $9.3 billion and imports of $7.3 billion (an import-export ratio of .12); the future looks grim for this industry. One indication of this is Japanese firms' domination of the lap-top market, one of the fastest growing segments of the computer industry.

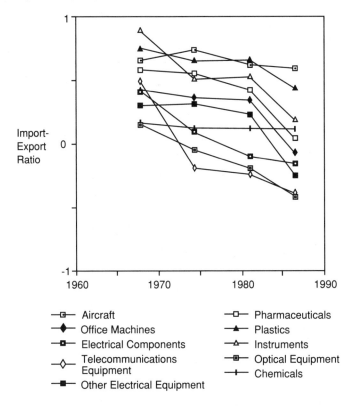

Figure 1-1. Import-Export Ratios for Specific Industries (U.S.)

Firms in the United States appear to be retaining their competitive strength in several high-technology industries, however. The import-export ratios for the aircraft, plastics, industrial and agricultural chemical, and pharmaceutical industries have declined very little over the last 20 years. These industries also still have positive trade balances; they have five of the top six trade surpluses among the high-technology industries. Further, these are the same industries in which Japanese firms are not doing well. Either Japan has a trade deficit in these industries or the industries' percent contribution to Japan's exports has decreased substantially over the last 20 years.

Reasons for Relative U.S. Strength in Aircraft, Chemicals, Plastics, and Pharmaceuticals

To understand the reasons for Japan's general success in manufacturing vis-à-vis the United States, it's useful to look in more detail at the aircraft, chemical, plastics, and pharmaceutical industries. Aircraft is the leading export industry in the United States. Most observers feel that U.S. defense spending, its large commercial market (geographically small countries don't need many aircraft), high barriers to entry (large investments are required), and the high-technology content of the aircraft industry (it has the highest R&D spending per sales dollar) have been the main reason for U.S. success in the aircraft industry. The only major non-U.S. producer of commercial aircraft, Airbus Industrie, required the cooperation of several European governments, which still heavily subsidize Airbus, to ensure its successful entry into the world market.[14] In other words, it's not clear that U.S. success or Japanese lack of success in the aircraft industry has much to do with either country's preferred style of management. Even if Japanese-style management is more applicable to the aircraft industry than U.S.-style management, it may take many years for Japan to overtake the United States in this industry.

The reasons for the success of U.S. firms in the other high-technology industries is more complicated. Many observers claim that since these industries are resource intensive, U.S. firms have an advantage over Japanese and European firms that do not have the necessary raw materials in their home countries. There are, however, several problems with this argument.

First, U.S. chemical and pharmaceutical firms have to purchase raw materials such as oil on the world market just like any other chemical or pharmaceutical firm. Japanese firms can purchase these raw materials from U.S. firms or from countries in the Middle East. After all, Japanese firms have historically performed well in industries for which the key raw materials were not available in Japan. For example, Japan's steel industry succeeded although the raw materials necessary to make steel are not abundant in Japan. Japanese firms merely built their factories on the coast, unloaded the raw materials directly from ships into the factories, and loaded the finished steel back onto the same ships. Japanese firms could use a similar strategy in the chemical and pharmaceutical industries.[15] Second, German chemical and pharmaceutical companies have succeeded and continue to succeed despite Germany's lack of appropriate raw materials. Germany has a positive import-export ratio in each of the high-technology chemical and pharmaceutical industries shown in Table 1-1 and Figure 1-1.[16]

The availability of the necessary raw materials does not seem to be an adequate reason for the continued success of U.S. firms and Japan's continued relative lack of success in these industries. There must be other reasons for this situation. One key reason put forth in this book is that Japanese-style management is less applicable to the nondiscrete parts industries — such as the chemical, plastics, and pharmaceutical industries — because they require less teamwork than is required in the discrete parts industries.

The Discrete Parts Industries

The discrete parts industries primarily produce products that contain parts. Traditionally these industries have been called the metalworking industries. However, because plastic parts are now commonly used, I use the term *discrete parts* to describe these industries. Using the U.S. Department of Commerce's system of Standard Industrial Classification (SIC), Figure 1-2 distinguishes discrete parts and nondiscrete parts industries.

Table 1-2. Discrete Parts and Nondiscrete Parts Industries

SIC	Industry	Discrete Parts Industry	Nondiscrete Parts Industry
20	Food products		x
21	Tobacco products		x
22	Textile mill products		x
23	Apparel products		x
24	Lumber products		x
25	Furniture & fixtures		x
26	Paper products		x
27	Printing & publishing		x
28	Chemical products		x
29	Petroleum products		x
30	Rubber products		x
31	Leather products		x
32	Stone, clay, and glass		x
33	Primary metal products	x	
34	Fabricated metal products	x	
35	Nonelectrical machinery	x	
36	Electrical machinery	x	
37	Transportation equipment	x	
38	Instruments	x	
39	Miscellaneous manufacturing	x	

Japanese firms have done well primarily in the discrete parts industries while U.S. firms are having more problems in those industries than in the nondiscrete parts industries. Of Japan's top 50 industries in world export share listed by Michael Porter, 48 are classified as discrete parts industries.[17] Table 1-1 demonstrates a similar situation. Since chemicals, plastics, and pharmaceuticals are all classified as part of the chemical industry (SIC 28), six of the seven discrete parts industries have seen their percentage contribution to Japan's exports increase over the last 20 years.

However, Japan is much weaker in the nondiscrete parts industries, particularly in the high-technology areas of those industries. It has a trade deficit in these industries, the exports from these industries as a percentage of total exports are decreasing, or both. Japan has a trade deficit in pharmaceuticals and in industrial and agricultural chemicals; the export contribution from its plastics industry has been dropping over the last 20 years. In fact, Porter states, Japan is strong in the chemical sector only when the products are inputs to discrete parts industries, particularly the auto and electronics industries.[18] As argued earlier, if the only difference between these high-technology industries was the level of technology, Japanese firms should be doing much better in the chemical, plastics, and pharmaceutical industries than in some of the electronics industries, since these nondiscrete parts industries require a lower level of technology than the electronics industries.

On the other hand, U.S. firms have been doing well in the nondiscrete parts industries. Of the top 50 U.S. industries in terms of world export share, 22 of the industries are classified as nondiscrete parts industries while only 16 are classified as discrete parts industries (12 of the industries are classified as agricultural or mining industries).[19] This is particularly true in the high-technology industries shown in Figure 1-1. The United States still has a substantial trade surplus in the pharmaceutical, plastics, and industrial and agricultural chemical industries. Although the import-export ratio has declined somewhat for these industries, they are still far higher than the ratios for the discrete parts industries.

This book is primarily about the high-growth discrete parts industries, in which Japanese firms have become the most competitive, and it argues that teamwork is particularly important in these industries.[20] Although many of the examples in this book are drawn from the semiconductor manufacturing equipment and the semiconductor industries, the issues, lessons, and conclusions discussed here apply to most of the discrete parts industries, primarily those industries that produce complex products containing many parts and requiring many manufacturing steps. The discrete parts industries shown in Figure 1-2 produce such products as engines, machinery, industrial equipment, appliances, consumer electronics, automobiles, ships, and musical instruments, in addition to the high-technology products shown in Table 1-1 and Figure 1-1.

CHAPTER TWO

A Comparison of Explanations for Japanese Manufacturing Success

NUMEROUS books have been written about Japan's economic success vis-à-vis the United States, particularly in the discrete parts industries. Due to the variety of approaches taken by the authors, it is often difficult to compare theories. Therefore, I will build on a framework adapted from William Abernathy, Kim Clark, and Alan Kantrow to classify the approaches taken by these books.[1] This framework, shown in Figure 2-1, divides approaches into four categories using a two-by-two matrix. One dimension of the matrix differentiates macro-level (country) and micro-level (company) issues, while the other distinguishes hard (policies/strategies) and soft (infrastructure/cultural) issues.

Figure 2-1 shows a number of issues in each quadrant of the matrix that have been raised by books attempting to explain Japan's economic success. The first quadrant deals with specific government policies. These policies include fiscal and monetary policies, regulations, and import-export laws. The second quadrant deals with a country's culture; it includes religion, social

	Structure	**Infrastructure**
Macro	**Government Policies** Ministry of International Trade and Industry Import restrictions Low-cost capital	**Social Infrastructure/Culture** Education Work ethic Group-oriented culture
Micro	**Corporate Policies** Just-in-time manufacturing Focused factory Product innovation Concurrent engineering High debt-to-equity ratios Fast introduction of new products Close relations with suppliers	**Corporate Infrastructure/Culture** Lifetime employment Seniority system Enterprise unions Nonspecialized career paths Collective decision making Collective responsibility Extensive training programs

Adapted from William Abernathy, Kim Clark, and Alan Kantrow, "The New Industrial Competition," *Harvard Business Review*, September-October 1981.

Figure 2-1. Reasons Given for Japan's Economic Success, Classified into Four Areas Related to a Country's Economic Competitiveness

mores, work ethic, education, and the way people are supposed to deal with each other. Chalmers Johnson, Clyde Prestowitz, Ezra Vogel, Edwin Reischauer, and Robert Reich have focused primarily on these quadrants in their explanations of Japanese economic success.[2]

The third quadrant focuses on specific corporate policies. Just-in-time (JIT) manufacturing, quality improvement, concurrent engineering, and the focused factory are some of the many corporate policies originally unique to Japanese firms. A number of books that discuss Japan's manufacturing success have addressed these issues; James Abegglen and George Stalk, Richard Schonberger, and Yasuhiro Monden are a few of these authors.[3] Most manufacturing managers seem to believe that the reasons for Japan's economic success lie in this quadrant.

The fourth quadrant deals with a firm's infrastructure or "culture." Lifetime employment, seniority system, enterprise unions, multifunctional as opposed to specialized workers, cross-functional job rotation, and collective responsibility are some of the issues that have been raised in this quadrant. William Ouchi, Richard Pascale and Anthony Athos, and Lester Thurow are among the authors who have addressed these issues.[4] Many of these issues are obviously related to some of the issues located in the second quadrant.

I believe that the sources of Japan's economic success lie in all of these quadrants, primarily due to the interdependencies between them. Japan's corporate policies and to some extent its government policies clearly provide Japanese firms with a competitive advantage. Corporate strategies such as JIT manufacturing, concurrent engineering, close relations with suppliers and customers, and even some of the policies of the Ministry of International Trade and Industry (MITI) have been shown to provide Japanese firms with a competitive advantage.

I also believe that several elements of Japan's corporate culture — some of which confused our hypothetical engineer in his first exposure to a Japanese company — support the implementation of these strategies due to the importance of teamwork to these strategies. Japan's corporate culture facilitates teamwork, and teamwork facilitates the implementation of Japan's unique corporate policies. Furthermore, several aspects of Japan's social infrastructure and culture obviously influence Japan's corporate culture. Before taking up this argument, however, let us review the various explanations given for Japanese economic success, using Figure 2-1 as a guide.

Japanese Government Policies

Several works have argued that Japan's economic success can be attributed to its government policies. These government policies can be divided into the policies promoted by MITI, import restrictions, and the low cost of capital.

Policies of the Ministry of International Trade and Industry

Chalmers Johnson has written what is probably the most comprehensive book about MITI. In *MITI and the Japanese Miracle*, he describes MITI's substantial effect on Japan's postwar economic growth. Through a combination of special benefits orchestrated by MITI, including low-interest loans, subsidies, tax benefits and protectionism, Johnson argues, Japan was able to develop a number of critical high-growth industries. In particular, MITI managed Japan's limited foreign exchange reserves, which is always a problem with developing countries, and it had access to Japan's postal savings. MITI used its control of foreign exchange to allow the importation of only those goods it felt were critical to developing export capability in a few key industries. It also used the low-interest funds from Japan's postal savings fund to finance the growth of these key industries.[5]

Prestowitz makes a similar argument in *Trading Places*, although in a broader context. He argues that Japan has focused on a few key industries and has emphasized cooperation among business, government, and labor to develop world-class capability in these industries. MITI is the primary, albeit not the only, conductor of this coordination, according to Prestowitz. MITI coordinates Japan's R&D policy and provides capital and licenses to firms in key industries. Although the funds were not particularly large, the licenses enabled the firms to import key technologies that could not otherwise have been imported.[6]

MITI's effect on Japan's economic success is controversial. Johnson demonstrates the importance of MITI to Japan's early postwar economic success. Through MITI's management of the Japanese foreign exchange, Japan was one of the few developing countries to avoid foreign exchange problems without excessive dependence on foreign capital.[7] Much of the early growth (1950s) occurred in a few key industries such as steel, commodity chemicals, and textiles, which were supported by low-interest loans from MITI.

By the 1970s and 1980s most of Japan's large companies were in existence and were doing quite well, particularly Japan's automobile and consumer electronics firms. Johnson and Prestowitz argue that MITI has enabled and continues to enable the Japanese government to target specific technologies and facilitate technology transfer between different firms. In addition, they argue, these cooperative research projects reduce the financial risk to Japanese firms, since multiple firms are involved in projects. These arguments are probably valid. Although the U.S. government funds more than twice the percentage of total research in the United States than the Japanese government funds in Japan (in 1986, 47 percent versus 20 percent), Japanese research seems to be more productive than U.S. research.

However, MITI's effect on Japan's economic success is debatable. There are two reasons why MITI cannot be the only reason for Japanese economic success and probably is not even the biggest reason. First, we have already seen that Japanese firms have not been that successful in the chemical, plastics, and pharmaceutical industries despite heavy support from MITI in the form of research funds and import protection from the Japanese government.[8] Second, MITI does not appear to be the only reason why Japanese firms are able to perform collaborative research. It appears that Japanese firms in general are better than U.S. firms at integrating research from multiple internal businesses and working with suppliers on research projects. The second argument is supported by the research of C. K. Prahalad and Gary Hamel; Michael Porter advances both arguments.[9] According to this thinking, U.S. firms need to learn at least as much about how Japanese firms perform research internally and work with suppliers as about how MITI orchestrates the projects.

Import Restrictions

Johnson, Prestowitz, and others also argue that MITI uses import restrictions to support key industries, and most economists seem to agree with this conclusion. If there was a free market, the prices of foreign products in Japan would be similar to the prices of these products in other countries. Studies have found that most foreign — and even Japanese products — are much more expensive in Japan than they are in the United States and European countries. Further, even as far back as 1967, Japan had a trade surplus in every high-technology industry shown in Table 1-1 except aircraft and pharmaceuticals. However, Japan was obviously not a world leader at that time in any of these industries. According to Johnson, Prestowitz, and Porter, Japan made a conscious decision not to import these products.[10]

The benefits of import restrictions are more debatable. In this context, it is not important to debate the relative advantages and disadvantages of restricting imports, however. What is important is that import restrictions are obviously not the driving force behind Japan's economic success. Many other countries in the world restrict imports (including the United States) and none of these countries has experienced the type of economic growth seen in Japan. In many countries (the United States, for example), industries that have received protection from imports have become less competitive after import restrictions were implemented. Import restrictions may provide Japanese firms with some breathing room before they enter the international marketplace. But once these firms do enter the international marketplace, well-run manufacturing systems, well-designed products, and new technologies become the means of competition. These new products, improved systems, and new technologies are not the direct or even the indirect result of import restrictions.

Low Cost of Capital

Japan's cost of capital is another aspect of Japanese government policies that receives a great deal of attention. Japan has a higher savings rate than the United States. Japanese manufacturing firms use higher debt-to-equity ratios than U.S. firms due to the close relations they have with Japanese banks.[11] According to this argument, the combination of a higher savings rate and lower debt-to-equity ratios enables Japanese firms to have access to lower cost capital than U.S. firms. This low-cost capital allows Japanese manufacturing firms to more easily invest in new technology and new capital equipment. Almost every discussion of Japan's economic success mentions low-cost capital as one reason. Economists in particular focus on this aspect. Several economists are so convinced that low-cost capital is the driving force behind Japanese economic success that they have predicted the downfall of Japan due to a rising cost of capital there.

However, it's not clear that low-cost capital is the most important reason for Japanese economic success. Researchers have questioned this argument from at least three points. First, not all economists agree that Japan has a significantly higher savings rate or a significantly lower cost of capital. For example, Fumio Hayashi argues that due to differences in the way Japan and the United States develop national income statistics, principally in the way capital is depreciated and government expenditures are categorized, Japan's savings rate has been only about 5 percent higher than U.S. savings rates over the last 15 years.[12] Robert Lawrence claims that Japanese companies float at least half of their bonds in international markets. He goes on to ask, "Why would they do that if they can borrow at lower rates in Japan?"[13]

A second problem with the cost of capital being the driver of Japanese economic success is that other factors can influence a firm's investment costs more than the cost of capital. Robert Hayes, Steven Wheelwright, and Kim Clark argue that a firm's level of work-in-process inventory and its product development time have a greater effect on a firm's investment costs than the cost of capital.[14] Almost every study has found that Japanese firms carry less inventory than U.S. firms and that it takes less time for a Japanese firm to develop a new product.

Third, no one has developed a correlation between the degree of Japanese success or U.S. lack of success in a specific industry and the capital intensity of that industry. If the cost of capital is so important, Japanese firms should be doing better in capital-intensive industries while U.S. firms should do well only in non–capital-intensive industries. Yet Japanese firms are not doing well in many capital-intensive industries, such as the chemicals, plastics, food processing, textiles, rubber, and paper industries, while U.S. firms are very competitive in these industries.

Social Infrastructure and Culture in Japan

The second quadrant in Figure 2-1 represents the cultural aspects of a country that affect economic competitiveness. These issues are the soft side of a country's national structure. They don't change easily or quickly, particularly not through national laws. They depend on a historical tradition and they evolve slowly. Education, the work ethic, and a group orientation are three aspects of Japanese culture frequently mentioned as having a positive impact on the country's economic growth.

Emphasis on Education

Almost every discussion of Japan's manufacturing success mentions the Japanese educational system as a primary reason for that success. Japan graduates a larger percentage of its students from high school than the United States does. More Japanese students can read and write when they graduate — Japan has a higher literacy rate than the United States despite its more complicated system of reading and writing. Thomas Rohlen found that Japanese students also score higher on standardized tests than do U.S. high school students and that a typical Japanese high school graduate knows as much as a college junior in the United States.[15] His conclusion is based on comparisons of the performance of Japanese and U.S. students on various tests and the number of hours that Japanese and U.S. students spend studying various subjects.[16]

Japan's superior system of high school education probably provides Japanese firms with better educated production workers than U.S. firms have access to. Japan's educational system enables these production workers to handle advanced manufacturing technologies and various statistical and mathematical techniques better than U.S. production workers. Japanese firms operating in the United States reportedly reject a larger percentage of U.S. applicants than Japanese applicants because the Americans cannot pass the proper tests.

For all of the apparent advantages that Japanese production workers seem to possess, however, few Americans want to adopt Japan's educational system in large part.[17] Most of Japan's high school and junior high school educational systems are geared toward college entrance examinations, multiple-choice tests that require students to regurgitate vast quantities of facts. Japanese parents try to enroll their children at junior high and high schools whose students have been the most successful in these examinations. Therefore, the good schools, along with evening cram schools (*juku*), basically teach children the facts they will need to perform well on the entrance examinations. High school students learn little about how to think or solve problems.

How many Americans believe this a vastly superior educational system — a system so much better than the U.S. educational system that it is driving Japan's manufacturing success?

Japan's universities receive even less praise from U.S. and Japanese observers.[18] Once students are accepted into a Japanese university, they are virtually guaranteed graduation. Many students rarely attend class; graduates recall their university days as filled with playing mah-jongg or shogi, two popular Japanese games. Another indication that Japanese universities are a large step below U.S. universities is the low level of cross-fertilization between Japanese universities; there is a great deal of inbreeding. Further, Japanese companies actually give less money to Japanese universities than to U.S. universities.[19]

Even if one accepted that Japan has a superior pre-university educational system, no one would argue for the superiority of Japan's university system in comparison with the U.S. system. Several studies suggest that engineers (i.e., university graduates) have a larger effect on a product's manufacturing costs than production workers (i.e., high-school graduates). More than 85 percent of a product's life cycle costs are determined before a manufacturing department even becomes involved with a product; more than 90 percent of the costs are determined before the production workers are involved.[20] To reduce a product's manufacturing costs, then, the education that an engineer receives is arguably more important than the education that a production worker receives. It's hard to believe that the advantage of well-educated production workers outweighs the burden of poorly trained engineers enough to argue that Japanese education is an important factor in Japan's economic success.

The Work Ethic

A second area of Japan's culture often thought to affect Japanese economic success is its strong work ethic. Japanese junior high and high school students work extremely hard. Even if many Japanese and Western observers believe Japanese students are learning many of the wrong things, Japanese students certainly spend more time in school, and they study more than U.S. students. In addition to the longer hours that Japanese schools are open each year as compared to U.S. schools, most Japanese students also attend *juku*, which meet in the afternoon and evening.

Students in the U.S. also work hard — in the local McDonald's or at some other part-time job. While few Japanese students have part-time jobs, most U.S. students work in the summer and on weekends during the school year to earn spending money. The culture in the United States encourages young people to have part-time jobs. We believe that these jobs teach kids good values, and they probably do. But these jobs take time away from school. It is difficult for students to work and still study.

Japanese adults also work harder than adults in the United States. In 1986, the average Japanese adult worked 212 hours more than the average American adult and 495 hours more than the average West German adult.[21] Many economists argue that this enables Japanese firms to have a lower hourly employee cost than U.S. and European firms. However, there are a number of reasons why this is probably not very important. First, Japanese employees are paid for most of their overtime, although this is very rare in a U.S. firm. Second, people in many countries of the world, particularly those from third-world countries, work longer hours than Americans merely to survive. In fact, there is probably a negative correlation between a country's level of income and the average number of hours its people work.

Most people would probably argue that the number of hours worked is not as important as how those hours are worked. Aren't Japanese employees working harder than U.S. employees? Aren't they working harder on the "right things"? Probably so, but these issues are related to the way an individual firm is managed. Whether employees are motivated or working on the right things obviously depends a great deal on how individual firms are managed.

A Group-oriented Culture

Japan's group-oriented culture is a third area often thought to affect Japan's manufacturing competitiveness. One example of this culture can be found in Japan's primary schools. In *The Japanese Educational Challenge,* Merry White describes how Japanese students learn to discover things and solve problems together as opposed to getting to an answer fast: "assignments are made to groups. Individual progress and achievement are closely monitored, but children are supported, praised, and allowed scope for trial and error within the group. A group is also competitively pitted against other groups; a group's success is each person's triumph."[22] White also notes that Japanese children are taught how to work together in primary school. According to White, children are primarily responsible for cleaning the schools and serving lunch: "The hot lunch is picked up by a team of children, while the rest arrange the desks to form group tables."[23]

Benjamin Duke makes a similar observation that schools create loyal workers by teaching the importance of cooperation and responsibility to peers. According to Duke, the classroom group occupies center stage in virtually every aspect of school life, whereas schools in the United States tend to encourage individualism and independent choice.[24]

While most people agree that Japan's culture and its schools are very group oriented, the effect of this on Japanese manufacturing success is more problematic. Pascale and Athos were among the first U.S. observers to argue that Japan's group-oriented culture has had a positive effect on Japan's economic success. They believe that Japanese firms have better internal communication, that their employees are more motivated, and that their employees can better handle interdependent activities because of factors directly related to Japan's group-oriented culture.[25]

According to Pascale and Athos, Japanese firms have better internal communication than Western firms because Japanese employees are more accustomed to indirect communication, vagueness, and ambiguity — items that have become more important to a firm's competitiveness due to the increasing complexity of modern corporations and products. They argue that Japanese employees are more motivated because Japan has developed one institution — the corporation — to handle the individual's spiritual and productive time, whereas the West has developed separate institutions (i.e., churches and companies) to handle each of these issues. Since Japanese employees expect to receive more of their "spiritual" needs from their place of employment, this encourages them to identify more with their firms and thus work harder.

Pascale and Athos also argue that Japanese employees can better handle interdependent activities due to better skills at maintaining peace and harmony (i.e., _wa_). These skills enable Japanese employees to avoid destructive personal confrontations when solving problems involving a large number of interdependencies between different parts of a firm.

Prestowitz also argues that Japan's group-oriented culture is a reason for Japan's manufacturing success. He believes that this orientation helps Japanese employees and firms cooperate better. He also argues that this culture encourages groups, both internal and external to a firm, to develop long-term relationships. It also discourages individuals from doing business with people from outside of the group, the most notable example being foreigners.[26]

There is a great deal of controversy concerning the effect of Japan's group-oriented culture on its economic success. First, a good description does not yet exist for what people actually do in a group-oriented culture versus an individualistic culture. For example, many people interpret the group-oriented culture argument as a suggestion that there is some type of "Japan, Inc." where all Japanese are working together to defeat the rest of the world. But as many people have pointed out, there is a great deal of economic competition in Japan. Second, a clearer connection needs to be made as to how Japan's group-oriented culture affects motivation, internal communication, and the management of interdependent activities, as well as between these three items and the country's economic success. For example, how do these factors lead to efficient manufacturing systems, well-designed products, and new technologies?

Japan's Corporate Culture

Pascale and Athos argue that Japan's group-oriented culture has caused Japanese firms to develop a unique corporate culture that is the prime reason for Japan's economic success. They developed a "7S framework" to help U.S. firms apply aspects of Japan's corporate culture to their operations and improve their competitiveness. Firms in the United States, in their opinion, focus only on three "hard" aspects of the 7S framework: strategy, structure, and systems. Japanese firms, they argue, are performing well because they focus more on the "soft" aspects of the 7S framework: staff, skills, style, and superordinate goals.[27]

Staff refers to the demographic makeup of a firm and *skills* refer to the capabilities of the employees. According to Pascale and Athos, Japanese firms develop better staff and skills through extensive training, particularly for new employees, and through cross-functional job rotation.[28]

Style refers to the cultural style of an organization and how the organization solves problems. Pascale and Athos believe that Japanese firms have a better style than U.S. firms because they develop problem-solving methods that take into account the interdependencies between people. Firms in the United States, they believe, tend to focus only on "output" and do not spend enough time improving their problem-solving methods.[29]

Superordinate goals refer to the significant guiding concepts that an organization imbues in its members. Japanese firms tend to take a more "holistic" approach with their employees; firms feel responsible for their employees' whole lives, not just their work time. Pascale and Athos argue that this approach encourages and causes Japanese employees themselves to be more holistic (e.g., to work toward corporate, as opposed to segmented, goals) in their approach to their jobs.[30]

Theory Z

William Ouchi is another researcher who maintains that Japan's corporate culture is an important reason for Japan's economic success. In *Theory Z*, he contrasts U.S. and Japanese management in seven specific areas (see Figure 2-2).[31] These areas revolve around Japan's group-oriented culture and its supposed emphasis on the success of the group (in this case a manufacturing firm) over the success of the individual.

With respect to lifetime employment, the group orientation encourages Japanese people to work for one firm as long as possible, often for their entire lives. Slow evaluation and promotion (sometimes called the seniority system) encourages employees to focus on the long-term health of the firm as opposed to short-term personal goals. Nonspecialized career paths (sometimes called job rotation) enable employees to learn about the entire firm and, according to Ouchi, help integrate different parts of a firm. Implicit control refers to a set of shared values that are

Japanese Organizations	American Organizations
Lifetime employment	Short-term employment
Slow evaluation and promotion	Rapid evaluation and promotion
Nonspecialized career paths	Specialized career paths
Implicit control mechanisms	Explicit control mechanisms
Collective decision making	Individual decision making
Collective responsibility	Individual responsibility
Wholistic concern	Segmented concern

Source: William Ouchi, *Theory Z: How American Business Can Meet the Japanese Challenge* (Reading, Mass.: Addison-Wesley, 1981), pp. 48-49.

Figure 2-2. Ouichi's Comparison of Organizational Models

used to manage the firm. Collective decision making, collective responsibilities, and holistic concern refer to how Japanese employees make decisions, share responsibility, and tend to be more broad in their outlook than U.S. employees. Ouchi argues that these aspects of corporate culture help Japanese firms integrate different sections, departments, or divisions of a firm.

Ouchi also argues that these seven aspects of Japanese management enable Japanese firms to improve productivity faster than U.S. firms through their positive effect on teamwork. He claims that "productivity is a problem that can be worked out through coordinating individual efforts in a productive manner and of giving employees the incentives to do so by taking a cooperative, long-range view."[32] Ouchi's book provides a few examples of this teamwork. He describes the positive effects of these seven aspects of Japanese management on employee-management relations, union-management relations, and supplier-customer relations. He argues that these seven aspects encourage Japanese employees to work hard and cooperate with employees from other departments and promote long-term relationships between suppliers and customers.[33]

Clarifying the Link Between Japan's Corporate Culture
and Economic Success

I agree with these arguments that Japan's corporate culture is a major reason for its economic success. We still need a more direct connection, however, between this corporate culture and the economic results. Three questions remain, and although this book does not claim to completely answer them, it takes us part of the way.

First, many of the elements of Japan's corporate culture are still not well understood. For example, how do the skills actually differ between U.S. and Japanese workers? How are the problem-solving styles of U.S. and Japanese firms different? What is the quantifiable difference between implicit versus explicit control mechanisms? How many people actually are involved with making a decision or sharing responsibility in a situation that has collective decision making and collective responsibility? How do we quantify a nonspecialized or specialized career path?

Second, how do the specific elements of Japan's corporate culture directly result in well-designed products, productive manufacturing systems, and new technology?

Third, the United States is a very successful country and its success seems to be at least indirectly attributable to individual effort, not group effort. Our economic history is filled with individuals who popular culture believes are primarily responsible for U.S. economic success. Yet many of the elements of Japan's corporate culture are a direct contradiction to "individualism." If the group-oriented culture is so important and so different from the emphasis on individualism, one wonders how the United States and Japan have both succeeded in the same world and in the same century.

Before beginning to address these issues, we need to take a look at other aspects of Japanese management that many

Japanese experts argue have a direct affect on Japan's economic success.

Corporate Policy in Japanese Companies

It's clear that many of the corporate policies listed in the fourth quadrant of Figure 2-1 have had a direct effect on Japan's economic success. These policies include just-in-time manufacturing, total quality control (TQC), incremental improvement of products and processes, the quick introduction of new products, concurrent engineering, the fast commercialization of new technology, and close relationships between banks, manufacturing firms, and suppliers. Judging by what is written in manufacturing journals and magazines, most U.S. middle and lower managers in manufacturing firms seem to believe that the issues in this quadrant are the most important reasons for Japan's economic success.

Japanese Manufacturing Strategies

In the late 1970s, many U.S. managers started visiting Japanese factories and finding results that contradicted Western beliefs about appropriate manufacturing strategies. These managers found that Japanese factories were carrying less inventory, achieving higher yields, and using fewer suppliers than most of the managers thought possible or even reasonable. In *Japanese Manufacturing Strategies*, Richard Schonberger describes how Japanese firms continually reduce inventory and improve quality primarily through just-in-time manufacturing and total quality control. JIT manufacturing is a continual improvement strategy that focuses on exposing and solving problems in order to reduce inventory and cycle times. Schonberger describes how Japanese factories improve quality and achieve significant reductions in delivery time, inventory cost, and direct and indirect labor costs through exposing and solving quality problems, reducing setup time and lot sizes, simplifying material flow and production systems (e.g., a pull system and mixed-model scheduling), and implementing preventive maintenance. Better working relationships with suppliers and customers is also an important aspect of these activities.[34]

Schonberger also describes how Japanese firms use total quality control to improve quality on the manufacturing floor. He argues that Japanese firms are able to obtain higher quality levels than U.S. firms by making everyone responsible for improving quality, particularly production workers. Japanese production workers inspect their own quality, stop the assembly line when they find quality problems, and use statistical process control to improve product quality.[35]

Incremental Improvement of Processes and Products

Since Schonberger's book was published in 1982, hundreds of books have been published in areas related to JIT manufacturing and total quality control. Many of these books have started to use the terms *continuous improvement* or *kaizen* in describing JIT manufacturing and total quality control. For example, in his book *Kaizen*, Masaaki Imai describes how Japanese firms use JIT manufacturing and total quality control to continually improve their factories.[36]

Many U.S. researchers have also found that Japanese firms are more likely than U.S. firms to apply a strategy of continuous improvement to their products. For example, Robert Hayes and Steven Wheelwright describe how Japanese firms emphasize frequent incremental improvements in products, whereas U.S. firms emphasize infrequent "strategic leaps."[37]

In *Made in America*, The MIT Commission on Industrial Productivity came to a similar conclusion, based on research performed by Edwin Mansfield and Ramchandran Jaikumar. Mansfield found that almost half of the R&D expenditures by

U.S. industry are aimed at entirely new products and processes, whereas Japanese firms allocated only one-third of these expenditures to new products and processes.[38] Jaikumar, in his study of flexible manufacturing systems, found that Japanese firms emphasized the continual improvement of these systems, whereas U.S. firms felt that "if it ain't broke, don't fix it."[39]

Closer Relationships with Suppliers

An important part of incremental improvement is a close relationship with suppliers. Schonberger argues that suppliers are an important part of JIT manufacturing and that the closer relationships that exist between Japanese firms and their suppliers enable Japanese firms to better implement JIT manufacturing. Porter also argues that Japanese firms have developed strong ties with their suppliers. According to Porter, "[Japanese companies] stress cooperative long-term relationships with buyers and suppliers instead of opportunism. The principal function of the *keiretsu* (groups of companies affiliated by shareholding) and *shita-uke* structures [networks of many small- and medium-sized subcontractors and suppliers] is to facilitate interchange among related companies."[40] The MIT Commission on Industrial Productivity also concluded that one of the common reasons for U.S. competitive decline in the machine tool, automobile, and steel industries was poor cooperation between customers and suppliers.[41]

Concurrent Engineering

Japanese firms have also developed product development processes that are superior to those used by U.S. firms. Kim Clark and Takahiro Fujimoto found that Japanese automobile manufacturers are able to take a new car from the conceptual stage to the point of market introduction in about two-thirds the elapsed time and with one-half the engineering hours. Another study found that Japanese firms have been able to introduce mechanical transmissions, heating, ventilation, and air-conditioning equipment faster than U.S. firms.[42]

One strategy used by Japanese firms to quickly introduce new products is concurrent engineering or overlapping problem solving.[43] Concurrent engineering is a product development strategy in which design engineers and manufacturing engineers work together to concurrently or simultaneously design the product and the manufacturing processes. Concurrent engineering not only enables a firm to introduce a product faster than conventional techniques, it also enables the firm to design a product that is more manufacturable and thus less costly by considering the manufacturing processes earlier in the product development cycle.

The importance of concurrent engineering also suggests that Japan's economic success is due to more than excellence on the manufacturing floor. As noted earlier, studies have found that more than 85 percent of a product's life cycle costs are determined before U.S. manufacturing departments typically become involved with developing a new product. Therefore, Japan's manufacturing success may be more related to the Japanese approaches to product development than to what they do on the factory floor. In addition, it suggests that countries cannot be strong in only manufacturing or engineering. As Stephen Cohen and John Zysman point out, just as countries and companies cannot succeed at R&D without excellence in manufacturing, manufacturing firms cannot succeed without an excellent product development capability.[44]

Faster Technology Development

It also appears that Japanese firms are able to commercialize new technology faster than U.S. firms. The success of Japanese firms in high-technology industries seems to confirm this hypothesis. Much of this capability is related to the ability of Japanese firms to quickly develop and incrementally improve products using concurrent engineering. However, while concurrent engineering enables Japanese firms to enhance or improve products quickly, Japanese firms also seem to be able to move new technology from the laboratory to the marketplace or factory faster than U.S. firms. For example, Japanese firms can introduce new memory chips faster than U.S. firms.

One way Japanese firms speed the commercialization of new technology is through closer ties between laboratories and divisions. For example, one auto industry observer argues that Japanese ceramics researchers work more closely with their automaker clients than do similar U.S. researchers. "While U.S. firms have set a research target of reducing the grain size of the ceramic powder, the Japanese know that what is important to the end customer is not grain size, but a product that can withstand so many degrees for 50 hours or something you can bang on 50 times a day."[45] Edwin Mansfield's research seems to corroborate this anecdotal evidence. He found that one-third of the Japanese R&D projects were based on suggestions either from external customers or from the firm's own production personnel (also users), whereas only one-sixth of U.S. projects were suggested by these sources.[46]

Another way in which Japanese firms are able to quickly commercialize new technology is through linkages between related businesses. Prahalad and Hamel argue that different businesses in a Japanese firm work together to develop *core competencies:* technologies that are fundamental to multiple businesses and are too expensive or long term for any one business to support by itself. They maintain that NEC, Canon, and Honda have grown faster than GTE, Xerox, and Chrysler due to their more effective use of core competencies.[47]

Porter makes a similar argument. He argues that close relationships between firms in related and supporting industries are among the four most important determinants of a country's success in a specific industry. According to Porter, "the presence in a nation of competitive industries that are related often leads to new competitive industries."[48] He argues that Japan's success in copiers, other office machines, photographic equipment, and telecommunications equipment was a strong factor in Japan's success with facsimile machines because those products covered all of the essential technologies important to facsimile.[49]

Prestowitz also argues that Japan's economic groups (*keiretsu*) and enterprise unions (*kigyo keiretsu*) have had a strong impact on Japan's economic success through their effect on these linkages. Economic groups are collections of firms that have equity interests in each other, have common banking affiliations, and buy products from each other. Within these economic groups are enterprise unions that consist of a large manufacturing firm and its suppliers and distributors. Prestowitz maintains that the economic groups and enterprise unions lend stability to planning and encourage long-term thinking through their emphasis on "sticking together." He also argues that these institutions reduce the risk of making long-term investments in new technology because they have a captive bank that provides low-interest loans, even when times are bad. They also reduce risk because the various firms provide a captive market for new products that other firms in the group have developed. These captive markets enable firms to work closely with customers to identify and solve the problems associated with the new technology.[50]

In the same vein, Charles Ferguson argues that Japan's economic groups and enterprise unions are a primary, if not the main, reason for Japan's economic success in the electronics industries. According to Ferguson, six large Japanese companies (Hitachi, Fujitsu, NEC, Toshiba, Mitsubishi, and Matsushita) control most of Japan's electronics business: 80 percent of semiconductor production, 60 percent of semiconductor consumption, 80 percent of computer production, 80 percent of telecommunications equipment production, and about 50 percent of consumer electronics production. "All have close linkages, including equity cross-ownership, with suppliers and affiliated industrial groups, banks, insurance companies, and trading houses. They have a long history of cooperation with MITI, Nippon Telephone & Telegraph (NTT), and each other." For example, according to Ferguson, "NEC's semiconductor operations are coordinated with its other businesses; NEC circuits are designed into its other electronic products, its semiconductor design systems use NEC mainframes and supercomputers, personnel are transferred between divisions, and NEC manufactures a number of semiconductor products exclusively for its own computers."[51]

The Benefits of Japanese Corporate Policies

Japan's corporate policies obviously have had a strong impact on its economic success. Numerous observers have documented the benefits of using these strategies in both Japan and the United States. JIT manufacturing, total quality control, close relationships with suppliers, concurrent engineering, and fast technology development are strategies widely accepted by manufacturing firms all over the world.[52] Most U.S. firms are actively implementing or attempting to implement these strategies. Judging by the documented economic benefits of these strategies and their widespread acceptance by Western companies, it appears that these strategies may be the direct reason for Japan's economic success.[53]

The close linkages between manufacturing firms and their banks, customers, and suppliers also seem to be an important reason for Japan's manufacturing success. The benefits of close relationships between customers and suppliers are well documented, and this strategy is becoming popular all over the world. In particular, many U.S. and European firms are developing close relationships with their customers and suppliers. The Malcolm Baldrige National Quality Award even includes among its criteria the existence of a close partnership with suppliers and customers that feeds improvements back into the operation.[54] However, there remain a number of questions concerning the hypothesis that these corporate policies are the primary reason for Japan's economic success.

Reasons for Japanese Success with Advanced Manufacturing Strategies

Japan's development and continued greater use of these strategies raises two questions. First, why did Japanese firms popularize these strategies before U.S. firms? Although some of these techniques were originally developed in the United States (e.g., Henry Ford's assembly line and some of the quality control techniques), Japanese firms were using these techniques on a wide scale before U.S. firms even agreed that they should emphasize them heavily. Was this accidental or did some characteristic of Japan's corporate culture facilitate the use of these strategies? Did this corporate culture help Japanese firms identify the benefits to be gained from these strategies and does it help Japanese firms implement them?

Second, if it is only these corporate policies, and nothing else, that drives Japan's economic success, why don't U.S. firms just adopt these strategies and become more competitive? These ideas have been in the literature at least since the late 1970s. Further, Hayes, Wheelwright, and Clark point out, George Hyde wrote a book in 1946 that advocated, among other things, concurrent engineering.[55] Yet only a minority of U.S. firms appear to actually use concurrent engineering or the other unique Japanese strategies listed in Figure 1-3. Although many U.S. firms express a great deal of interest in these strategies, most studies have found that the understanding and implementation of these strategies are proceeding very slowly.[56] As Porter maintains, "A new theory must explain why firms from a particular nation choose better strategies than those from other nations."[57]

How Corporate Culture Supports Japanese Corporate Policies

One reason for the widespread use of these strategies might be Japan's corporate culture. Most observers of Japan have noted that corporate culture helps Japanese firms implement Japan's unique corporate policies. The elements of this culture sound similar to some of the aspects described by Pascale and Athos and by Ouchi. Schonberger, for example, refers to a number of elements of Japan's corporate culture in his books on Japanese manufacturing. In *Japanese Manufacturing Techniques*, Schonberger maintains that the use of multifunctional workers by Japanese firms helps them implement JIT manufacturing. He also argues that close relations between a firm and its suppliers are needed in order to implement JIT manufacturing.[58] In a later work he notes that a different form of organization, called the "customer-focused organization," facilitates the implementation of JIT manufacturing. Instead of the functional (e.g., engineering, manufacturing, marketing) organization typically used by U.S. firms, the customer-focused organization is organized into teams that focus on a specific product or a specific customer.[59]

Many of Schonberger's remarks suggest that Japan's corporate culture has some role in the implementation of these corporate policies. In particular, he seems to be advocating some of the elements described by Ouchi in *Theory Z*, among them a multifunctional work force to implement JIT manufacturing. The strategy to have close relations with suppliers sounds similar to Ouchi's terms "collective responsibility" and "collective decision making," in that the customer and supplier must share responsibility for a great number of activities. The customer-focused organization also implies the use of cross-functional job rotation and multifunctional workers (both advocated by Ouchi) within the context of a specific product.

Abegglen and Stalk also imply that several elements of Japan's corporate culture help Japanese firms introduce new products more quickly. They argue that U.S. firms have trouble building effective communication between different functional departments during the development of a new product because they use highly vertical organizations. In their view, each department in a U.S. firm typically takes its own parochial view during the development of a new product, which tends to make the product suboptimal from a systems point of view. They hypothesize that job rotation and the bonus system may help Japanese employees take a less narrow view.[60] The fact that Japan's organizations are less vertical than U.S. organizations seems to be culturally oriented. Organizational structures reflect a firm's or even a culture's ideas about responsibilities and the way these responsibilities are divided and shared. Job rotation and bonus systems, which Abegglen and Stalk hypothesize may be important to new product development, are also part of Ouchi's description of Japan's corporate culture.

The pervasive linkages between Japanese banks, manufacturing firms, suppliers, customers, and related businesses also appear to be culturally based. East Asian scholars have noted for years that Japanese people have a tendency to form groups more often than do Westerners.[61] Prestowitz argues that "just as individual Japanese do, Japanese companies tend to form groups." He argues that in Japan there "is an inevitable tendency toward alliance among companies. This led to the formation of the four great *zaibatsu* (large conglomerates of interrelated industrial, financial, and commercial enterprises)." After the end of the occupation in 1952, he argues, the *zaibatsu* were reassembled into their present-day form as *keiretsu* (economic groups).[62]

These linkages are also similar to Ouchi's definition of Japan's corporate culture. Under Ouchi's definition, Japan's corporate culture would seem to facilitate these linkages. For example, since cooperation between a Japanese firm and its suppliers requires some form of shared responsibilities and shared decision making, Ouchi's concepts of "collective responsibilities" and "collective decision making" would be applicable and beneficial to these linkages. Since these firms also need to take a more macro-level view of their business to make these long-term relationships work, Ouchi's concept of "holistic concern" also seems applicable.

Teamwork: A Link Between Japanese Corporate Policies and Culture

The point of this book is that there is indeed a relationship between Japan's corporate culture and the successful use of specific corporate policies by Japanese firms. The relationship between my concept of Japan's corporate culture, teamwork, its corporate policies, and its economic growth is diagrammed in Figure 2-3. Each of Japan's corporate policies requires teamwork, and Japan's corporate culture supports teamwork. By supporting teamwork, the corporate culture helps Japanese firms identify the potential benefits of these corporate policies. For the same reason, the corporate culture helps Japanese firms continue to implement their somewhat unique corporate policies more successfully than can be done in U.S. firms.

Teamwork is defined as extensive cooperation between employees in a business process. In this book, a business process is defined as a collection of activities that are performed by a number of different employees on a specific product or technology. Since these processes involve multiple people and multiple business functions (e.g., engineering, manufacturing, marketing), teamwork is an important part of these processes.

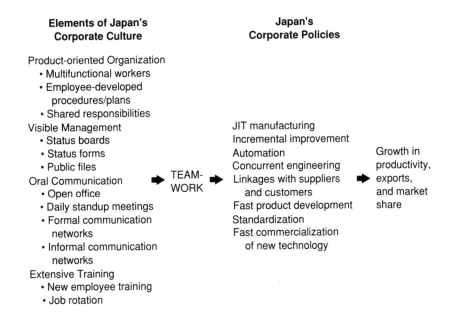

Figure 2-3. The Importance of Teamwork: The Relationship Between Japan's Corporate Culture, Corporate Policies, and Economic Success

Teamwork as a Major Element of Corporate Policies

Teamwork is needed to implement all of Japan's unique corporate policies. JIT manufacturing requires teamwork to reduce the amount of inventory that exists between each of the manufacturing steps. Almost everyone employed in a factory must be involved in these efforts: production workers, manufacturing engineers, purchasing and production control personnel, and management. Further, JIT manufacturing also requires extensive teamwork between a firm and its suppliers to reduce the amount of raw material inventory that exists in the customer's factory and the amount of finished inventory that exists in the supplier's factory.

Japan's other corporate policies also require extensive amounts of teamwork. Improvements in quality require teamwork because important quality problems tend to involve design engineering, multiple manufacturing processes, and suppliers. Concurrent engineering requires teamwork between employees from different functional departments such as engineering, manufacturing, and marketing. The achievement of low-risk, high debt-to-equity ratios requires teamwork between manufacturing firms and banks.

One study suggests that Japanese firms are more successful in those industries that require more teamwork. It compared the number of people needed to produce a variety of products in both U.S. and Japanese firms. In addition to finding that Japanese firms were able to produce all of the products with less labor content than the U.S. firms, it found that the productivity advantage of the Japanese firms was a function of the number of manufacturing steps needed to produce the product. When only 10 steps were needed to produce a product, Japanese firms had only a slight productivity advantage. When 100 steps were required, however, the Japanese firms had a two-to-one productivity advantage.[63]

It is easy to visualize that teamwork increases in importance as the number of manufacturing steps increases. More coordination is needed between the different manufacturing processes and the different functions in the factory when more manufacturing steps are required to make a specific product. In addition, since the number of parts in a product is often related to the number of manufacturing processes used to make a product, the amount of teamwork needed to design a product increases as the number of manufacturing steps increases.

Part Two describes in more detail how teamwork has become more important in those industries in which Japanese firms have been the most successful. It argues that teamwork has increased in importance over the last 70 years in the discrete parts manufacturing industries, the very industries in which Japanese firms have been the most successful. By focusing on three macro-level processes (production, product development, technology development) that exist in most discrete parts manufacturing firms, Part II describes how increased product and process complexity and shorter product lives have caused teamwork to become more important in these industries. It argues that the management style developed by U.S. firms in the early twentieth century is no longer adequate for today's complex products and processes and shorter product lives. In short, manufacturing — not U.S. firms or workers — has had the largest changes over the last 70 years, and Japanese-style teamwork is better suited to the current realities of discrete parts manufacturing.

How Japan's Corporate Culture Supports Teamwork

Part Three describes each of the elements of Japan's corporate culture shown in Figure 2-3. It uses examples primarily from one Japanese firm — Mitsubishi Electric Corporation — although it also refers to my experiences at Yokogawa Electric and visits to several other companies. In addition, it refers to numerous other studies including the books by Pascale and Athos and by Ouchi.

I would not argue that all or even most Japanese firms fit the description in Part Three. I believe that there are some differences between and probably within Japanese firms with respect to these elements of corporate culture. Further research is needed to illuminate these differences. I do note, however, that the differences between and within Japanese firms with respect to these elements of corporate culture are probably far less than the differences between the typical Japanese and the typical U.S. firm. After finishing Part Three, most readers will probably agree, particularly if they have experience in or knowledge of the way organizations operate in the United States.

Parts Four, Five, and Six show how these elements of Japan's corporate culture help Japanese firms manage the three macro-level processes described in Part Two. In Part Four, Mitsubishi's semiconductor factories are used as the vehicle to describe how the elements of Japan's corporate culture help Japanese firms implement JIT manufacturing and automation. In Part Five, examples from Mitsubishi's semiconductor equipment department demonstrate how the elements of Japan's corporate culture help Japanese firms develop high-quality and low-cost products in a short period of time. In Part Six, Mitsubishi's semiconductor business is used to describe how the elements of Japan's corporate culture help Japanese firms move technology from the laboratories to the factories.

Again, all Japanese firms are not the same as the examples described in Parts Four, Five and Six. Japanese firms vary in their emphasis on and implementation of JIT manufacturing, automation, concurrent engineering, close relationships with suppliers and customers, standardization, relationships between related businesses, and relationships between factories and laboratories. These strategies are affected somewhat differently by Japan's corporate culture. Given that Japanese firms have a different corporate culture and have been more successful than U.S. firms in implementing these unique strategies and policies, the concrete examples in this book suggest that a similar dynamic is probably taking place in many Japanese firms.

Part Seven describes not only how these elements of Japan's corporate culture can be applied in U.S. firms but also several potential problems with Japan's corporate culture and group-oriented culture. I have no intention of suggesting that Japan is perfect or that the United States should attempt to copy it. The issues are far too complex and there are too many areas in which I feel the United States is far superior to Japan. Even if the answers were clear, cultures are not changed easily, and I suspect that the United States will have a hard time learning from Japan. My point is to describe the potential problems with adopting aspects of Japan's corporate culture, noting at the same time that many of the things that U.S. industry must do are culturally related.

PART TWO

The Increasing Importance of Teamwork

Teamwork in the manufacturing industries has increased significantly in importance over the last 70 years. Increases in product and process complexity and shorter product lives have been the major reasons for this shift. The number of parts, the number of models, the complexity of individual parts, and the technological complexity of assembled products have all grown dramatically within this period. Further, since product lives are becoming shorter, the increased product and process complexity must be handled in shorter time frames. These changes are particularly relevant in the discrete parts industries in which hundreds or thousands of individual parts must be designed, fabricated, handled, and assembled.

Increased product and process complexity suggests the need for a management style different from the system developed by U.S. manufacturing firms in the beginning of the twentieth century. Based primarily on the teachings of Frederick Taylor, this system typically depended on a few experts to run it. Although this type of management style worked for U.S. firms in an earlier era, today's systems cannot be managed by a few experts without the involvement of virtually all employees. The complexity of modern products and processes require that production workers, support staff, design engineers, and research personnel all work together to improve product and process performance.

First, I will discuss those factors that are causing the increase in the importance of teamwork — how products and processes have become more complex and product lives have become shorter. I use the 1920s as a point of comparison because a number of critical changes occurred in manufacturing during that decade. During that time, the assembly line was popularized by Henry Ford, and General Motors introduced product diversity as a marketing strategy.

Second, by focusing on three macro-level processes that exist in all manufacturing firms, I will describe how these factors increase the importance of teamwork in manufacturing firms. In this context, a process is defined as a collection of multifunctional activities that are performed to produce a specific product (production process), develop a new product (new product development process), or develop a new technology (technology development process). Some people might think of these three processes as functions. Since they actually involve multiple functions, however, I use the term *process*. This is an important distinction, because U.S. firms organize themselves primarily by function, and employees are typically specialized in terms of function.

A firm's financial performance is largely dependent on how these three processes are performed. Most of a firm's business and manufacturing processes are merely subsets of these three macro-level processes. Cost, cycle time, and quality are three of the most common measures of process performance and many researchers have found a correlation between them. Short cycle time, low cost, and high quality have been found to be strongly correlated.[1]

Further, the activities performed by a firm produce value primarily when the firm completes one of the three processes. Firms typically receive payment after they deliver a product — when they complete the production process. New products become of value to a firm when they are being sold on the market — after the product design process is completed. New technologies become of value to a firm after they become part of a product or the processes (e.g., new equipment) that are used to produce products. Therefore, unless a firm's activities are viewed in terms of the entire production, product design, or technology development process, the performance of these activities is largely irrelevant.

CHAPTER THREE

The Changing Nature of Discrete Parts Manufacturing

S EVERAL changes have occurred since the 1920s that increase the importance of teamwork to the performance of the production, new product development, and technology development processes. Most of these factors are related to increased product and process complexity and shorter product lives; they include:

- Increased numbers of parts in a product
- Increased product diversity
- Increased compenent complexity
- Increased automation
- Increased technological complexity of products
- Shortened product lives

These six factors primarily affect the discrete parts industries (see Figure 1-2) because an increased number of parts, increased product diversity, and increased component complexity primarily affect those products that contain parts. Although increased automation, increased technological complexity of products, and shorter product life cycles also affect the nondiscrete parts industries, these factors have a much stronger effect on the discrete parts industries. Therefore, the importance of team-work has increased much more rapidly in the discrete parts industries than in the nondiscrete parts industries. Since Japanese firms have a corporate culture that supports team-work, they have a stronger competitive advantage inthe discrete parts industries than in the nondiscrete parts industries.

Increased Number of Parts

The typical number of parts in a product has grown considerably during this century. For example, the number of parts in an automobile has increased from a few thousand in 1900 to over 35,000 in 1980. The number of parts in an airplane has increased from about 100,000 in 1940 to over one million in 1980.[1] Highly complex products are also becoming a more important part of the manufacturing economy. Since high-technology industries tend to have higher growth rates than other manufacturing industries and tend to produce products that have more parts than those produced in lower technology industries, products with a large number of parts are becoming a more significant portion of the manufacturing economy. Consider the high-technology industries that were classified as a discrete parts industry in Part One. Aircraft, computers, telecommunications apparatus, and instruments typically have between 10,000 and 10,000,000 parts in a single product.[2]

Increased Product Variation

Seventy years ago, most products were produced with only a few options. Today, most products are customized for a specific order or produced with an endless number of options. For example, Westinghouse makes 50,000 different steam turbine blade shapes. AMP produces 80,000 different connector models. IBM's Selectric typewriter (with 2,700 parts) was produced in 55,000 different models.[3]

Increased Component Complexity

New manufacturing processes, such as centrifugal casting, injection molding, and isostatic powder metallurgy, enable firms to make more complex parts. Although typical machined parts may have only three to five distinct surfaces, these new fabrication techniques enable parts to have more than ten distinct surfaces.[4] Integrated circuits are probably the most complex components made today. They require more than 200 process steps, many of which involve the deposition of materials and the etching of complex patterns in these materials. If the sides and tops of each of these patterns are counted as distinct surfaces, today's memory chips probably have more distinct surfaces than could be found in all of the parts in automobile made in 1900.[5]

Increased Automation

The use of automation has also increased in the discrete parts industries. Computer-controlled machine tools, robots, and automatic assembly machines have been used only since the late 1950s. Automation is also being used to integrate individual pieces of equipment with business processes such as order entry, sales collection, and production control. Computer-aided design, computer-aided engineering, and computer-integrated manufacturing are just a few examples of how automation is being applied to white-collar jobs in manufacturing firms.

One result of increased automation is that direct labor costs have become a very small percentage of product costs. Jeffrey Miller and Thomas Vollmann found that direct labor costs in manufacturing industries decreased from 45 percent of value added in 1915 to less than 20 percent in 1980. Most of a product's costs (other than purchased materials) are now classified as overhead and are primarily associated with the movement of materials and the quality of these materials.[6] Several of the other factors mentioned above have also contributed to the increasing importance of this overhead. An increased number of parts and increased product variation have increased the relative contribution of overhead costs, since these two factors require a factory to receive, store, count, control, and handle a greater number of parts, particularly unique parts.

Increased Technological Complexity

A greater number of technologies affect an individual product than in the 1920s. Seventy years ago, products were either mechanical or electrical in nature. Today's products combine a multitude of mechanical and electrical technologies in new and different ways. The term *mechatronics* is often used to describe this new mix of technologies.

Shorter Product Lives

Seventy years ago, most products went for years without a design change. Today, products are continually improved, and new products are continually developed, primarily due to the increased speed of technological change. Even in the last 20 years, product lives have become shorter. A 1981 survey of 700 U.S. companies estimated that new products would account for one-third of all profits in the 1980s, an increase from one-fifth in the 1970s.[7] Another study found that in 1986 only 15 percent of the products studied had been introduced more than 10 years earlier, compared to 27 percent in 1975. The importance of new products will probably increase in the 1990s since many of today's products are modified each year, particularly high-technology products that often have a significant redesign each year. Due to shorter product lives, the time it takes to introduce a new product has become an increasingly important factor in a firm's success.[8] Therefore, not only have products and processes become more complex, but they need to be revised more frequently.

CHAPTER FOUR

The Growing Importance of Teamwork
in the Production Process

THE DRAMATIC changes in discrete parts manufacturing have made teamwork more important to the performance of the production process. The production process is an example of Michael Porter's concept of the *value chain*. Firms translate inputs from suppliers into outputs for customers. "The ultimate value a firm creates is measured by the amount buyers are willing to pay for its product or service. A firm is profitable if this value exceeds the collective cost of performing all of the required activities."[1] One firm's production process occupies a place in a long chain of raw material producers, component or subassembly manufacturers, final product manufacturers, distribution centers and warehouses, and finally customers or retail outlets. Nevins and Whitney call this the supplier-manufacturer-seller chain,[2] and it has become longer due to the increasing complexity of individual components. More complex components mean more levels of raw material and component suppliers. Each firm in this chain has its own production process, and these production processes are highly interdependent.[3]

Changes in the Production Process

A greater number of parts, greater product diversity, greater component complexity, and greater automation have increased the importance of teamwork to production performance, particularly in the discrete parts industries. Because each individual part typically requires several fabrication and assembly steps, a greater number of parts means that a greater number of manufacturing steps must be coordinated to produce a finished product. Greater product diversity means that there is a larger number of unique parts and a larger variety of process flows needed to produce all of the models in a product line. Each of these parts and assemblies may take a different path through the factory as part of its production process. It is easy for factories to become extremely complicated and unwieldy as product variation increases.

Greater Component Complexity

Greater component complexity also increases the importance of teamwork in the production process in at least three ways. First, complex components typically require complex manufacturing processes, which means that specialized manufacturing engineers are needed. Since specialists tend to focus on their specific area of knowledge, they often know little about the production system as a whole. As Robert Hall says, "Specialists may not be able to comprehend [the system], not because they are intellectually impoverished, but because whenever concepts that do not seem pertinent to their specialty are discussed, they lose interest in someone else's business."[4] The need to coordinate a greater number of these manufacturing specialists increases the need for teamwork in the production process.

Second, the complex manufacturing processes used to produce the complex components tend to be more interdependent than are less complex manufacturing processes. Complicated parts often require tight tolerances for flawless assembly. A slight variation in one fabrication process may make it difficult to assemble the parts. However, the source of the problem is not always clear, since the problem typically can be due to one of many manufacturing processes. Since several specialists often must work together to solve such a problem, greater component complexity also tends to increase the importance of teamwork in the production process.

Third, greater component complexity also increases the importance of suppliers and of teamwork between a firm and its suppliers. It is often more cost-effective to have suppliers produce special purpose components than to produce them internally. It appears that most manufacturing firms are having suppliers produce more of these individual components, since material costs are becoming a larger percentage of product costs, particularly in the electronics industries.[5] Suppliers are often better equipped with the special equipment and personnel needed to produce high-technology components such as castings, moldings, and integrated circuits.

Greater Automation

Greater automation also increases the importance of teamwork in the production process. Along with a greater number of parts and more product diversity, it has caused overhead to replace direct labor as the largest contributor to product costs. The declining contribution of direct labor costs has a tremendous effect on what is needed to be competitive in the production process. No longer can a manufacturing firm improve efficiency in only one part of a factory, because much of a factory's overhead is associated with the overall production system. Miller and Vollmann found that a large percentage of these overhead costs are due to system-related activities like material and quality control, particularly to the information or transactions associated with these activities. Departments such as production control, quality control, materials, warehousing, receiving, shipping, and management information systems contain many of the employees who perform these activities and incur these overhead costs.[6] Reducing this overhead requires teamwork among all of these departments.

The Need for System-level Improvements

Because most costs are at a system level, most meaningful improvements can no longer be made independently by one functional department or part of a factory. Improvements must occur at the system level. Given this argument, it's not surprising that successful factories perform well primarily at the system level. According to Hayes, Wheelwright, and Clark, "We were struck by the way all these elements seemed to mesh together in the highest performing factories. It was not so much excellence in particular areas of operation that distinguished [these factories] but the power of their manufacturing operations as a whole."[7]

Two ways to improve system performance are JIT manufacturing and automation/computer-integrated manufacturing. JIT manufacturing provides a methodology for improving a manufacturing system's performance by focusing on the reduction of cycle time through the production process and the reduction of inventory in the production process. Continuous flow manufacturing, small lot sizes, shorter setup times, kanban cards, mixed-model production, preventive maintenance, zero defects, and a minimum number of suppliers are some of the methods used to reduce cycle time in production.[8] Sometimes called a continual improvement philosophy, JIT manufacturing provides an approach to reducing inventory, identifying problems, and solving problems. It is a never-ending cycle of activities (thus the term continual improvement) in which a firm continually reduces cycle time and inventory.

The Importance of Teamwork to JIT Manufacturing

The most important part of JIT manufacturing, however, is people and teamwork. As Robert Hall puts it, "People make it happen."[9] Hayes, Wheelwright, and Clark maintain that, "when it comes to achieving sustained improvement in performance, the people in the organization and the way they are linked to the other elements of manufacturing should become the focus of attention."[10] Schonberger argues that "there must be massive involvement in the minute-to-minute problems that operators face on the shop floor."[11] Bruce Henderson observes that "no one knows better what the problems are and what needs to be done at each specific work station than the worker on the shop floor."[12]

Teamwork is an essential part of these activities since, according to Hayes, Wheelwright, and Clark, "most problems involve more than a single worker."[13] Schonberger believes that one of the greatest benefits of developing a continuous flow layout (one element of JIT manufacturing) is that it gets people into teams.[14] He becomes more adamant about teamwork in a later work, where he declares that "employees need to be organized into customer- or product-focused teams."[15]

Hayes, Wheelwright, and Clark believe that improved system performance and JIT require involvement from everyone. They maintain that employees need to "adopt a holistic perspective. We have repeatedly emphasized how different kinds of manufacturing decisions interrelate and reinforce one another." They also argue that "effective management depends much more on shared values — a common philosophy — than on superb analytical techniques."[16]

The most fundamental argument for teamwork comes from Richard Walleigh, who describes the common excuses put forward by functional departments as to why JIT won't work in their factories. Accountants, schedulers, sales staff, production controllers, and manufacturing engineers each have different reasons for why JIT won't work — and as Walleigh describes, each of these functional disciplines can derail a JIT effort.[17] Therefore, each discipline must be convinced that JIT will improve the factory's performance, and the argument must be made in a common language that each discipline can understand.

A common language, shared values, a holistic perspective, product-focused teams, involvement in shop-floor activities, and "people make it happen" are all different ways of saying that teamwork is important. No one said this 70 years ago, but people are saying it today because manufacturing has changed. Products have more parts and options, components are more complex, and overhead has replaced direct labor as the major driver of product cost. These changes in manufacturing increase the importance of JIT manufacturing and teamwork.[18]

JIT Manufacturing Requires More Teamwork in the Discrete Parts Industries

It is easy to see why Japanese firms have not been as successful in nondiscrete parts industries as in discrete parts industries. Since the nondiscrete parts industries make products that don't contain parts, their factories don't have as many manufacturing steps as factories in the discrete parts industries. The more manufacturing steps (due to a large number of parts or a great deal of product diversity), the more important teamwork becomes to the implementation of JIT manufacturing. Nondiscrete parts industries such as textiles, printing, lumber, paper, pharmaceuticals, and rubber have very few manufacturing steps. Japanese firms succeed in these industries only when they produce a major input to a discrete parts industry (e.g., rubber tires for the automobile industry).[19]

Some nondiscrete parts industries, particularly chemicals and plastics, have very few applications for JIT manufacturing except in transfers between plants. Large chemical and plastics plants do not have discrete manufacturing steps that can be moved around for the purposes of continual improvement. These plants are a series of pipes that transport various gases and liquids. They are typically built and operated to original specifications until they outlive their usefulness. The Japanese team-oriented management style, in which improvements are continually made to a factory's production process, is not as applicable to these types of products as it is to the products produced in the discrete parts industries.

Automation, Computer-integrated Manufacturing, and Teamwork

Automation and computer-integrated manufacturing (CIM) are also effective ways to improve the performance of the production system and thus the production process. However, it is clear that automation should only occur after the JIT-related improvements have been implemented. As Robert Hall states, "Much of automation's desired effect comes from advanced preparation and the discipline to use it wisely once in place. . . . If the benefit can be obtained from preparation and discipline, perhaps it can be obtained without investing much in the equipment and software itself."[20]

A similar view is found from the people who actually implement CIM. Charles Savage, a CIM consultant, maintains that "The interplay between [JIT, total quality assurance, and CIM] is critical, because JIT and total quality assurance help expose the real issues CIM must address. Without these three, CIM often amounts to little more than a computerization of the company's contradictions, confusion, and inconsistencies."[21] In other words, the manual system must be simplified before the computers are applied. JIT manufacturing provides a number of methods for simplifying the system, and as we found in the last section, teamwork is a large part of JIT manufacturing.

Teamwork is also important to the implementation of automation and CIM. Richard Walton argues that "The integration no longer makes it possible to define jobs individually or measure individual performance. It requires a collection of people to manage a segment of technology and perform as a team."[22] The people who implement CIM make similar arguments. Savage states that "although the challenges of CIM were once viewed as 80 percent technical and 20 percent organizational, practitioners now agree that clearly the reverse is true." He bases his views on surveys performed by the Yankee Group; Booz, Allen & Hamilton; the National Research Council; and the Automation Forum. In Savage's own survey of CIM users, respondents said that the greatest managerial change needed to implement CIM effectively was a greater emphasis on teamwork.[23]

CHAPTER FIVE

The Growing Importance of Teamwork in the New Product Development Process

T HE NEW product development process includes the activities needed to design a product, its manufacturing processes, and the logistical system needed to deliver the products to customers. It includes the definition of customer requirements, conceptual design, detailed design, and product testing and verification. Marketing, engineering, manufacturing, service, and other functions are involved with this process.

It will be easy to see why Japanese firms are more successful in the discrete parts industries than in the nondiscrete parts industries. The team-oriented approach to product development is much more important to the development of complex products that involve numerous parts, models, complex components, and many different technologies. This is characteristic of the discrete parts industries, particularly automobiles, consumer electronics, machinery, instruments, telecommunications equipment, semiconductors, and computers. On the other hand, it is not so characteristic of the nondiscrete parts industries, whose products have fewer parts, models, complex components, and technologies; the food, textile, lumber, paper, printing, rubber, plastics, pharmaceutical, and chemical industries are examples. Moreover, these industries do not require fewer product designers than the discrete parts industries. While a product designer is needed to design products that contain discrete parts, because of the logistical simplicity of products that have no parts, customers work directly with process engineers in the nondiscrete parts industries.[1] This significantly reduces the number of people and thus the teamwork needed to develop new products in these industries.

The Changing Nature of Manufacturing

All of the factors shown in Figure 3-1 — greater number of parts, greater product diversity, greater component complexity, greater technological complexity in products, greater automation, and shorter product lives — increase the importance of teamwork in the new product development process. A greater number of parts and a greater amount of diversity in a specific product increases the number of people needed to design the product. Greater component complexity and a product's greater technological complexity also increase the number of people involved because more technological specialists are needed. The greater the number of engineers and other employees involved with the design, the greater the importance of teamwork.

Greater component complexity also increases the importance of teamwork between a manufacturing firm and its internal or external suppliers, since complex components require more levels in the supplier-manufacturer-seller chain. Teamwork has always been important between a firm and its suppliers. Michael Porter maintains that effective supplier-customer relationships are one of the most important factors determining a country's economic success in a specific industry, as suppliers help firms identify new technologies and ways to apply these technologies.[2] Suppliers, therefore, need to be part of a firm's new product development process. However, greater component complexity requires the involvement of a larger number of suppliers than 70 years ago.

Greater automation and shorter product lives also increase the importance of teamwork. Increased automation means that fewer problems can be solved after the product has been designed since automated processes are less flexible than manual processes. Shorter product lives mean that the design activities must be performed in a shorter time frame. Products can no longer be "redesigned" for manufacturability or other reasons after they have been introduced.

These factors increase the importance of teamwork in a process that has always been considered highly interactive. Americans have recognized for years that the design engineering department must work with the manufacturing department to design products and processes simultaneously.[3] More than 85 percent of manufacturing costs typically are committed during the conceptual design of a product.[4] Without extensive interactions between the design engineering and manufacturing departments, these costs cannot receive adequate consideration.

The use of standard components and assemblies in multiple models or product lines to reduce product costs also requires a great deal of teamwork. To define the most appropriate standard components and assemblies, there must be a constant give-and-take between what is needed by the individual market segments (the marketing department), the process flows that exist in the factories (the manufacturing department), and the technological capabilities of the product (the engineering department).

Large Organizations Make Teamwork More Difficult

Although most observers agree that U.S. firms have evolved from a team-oriented approach to a sequential approach, the reason for this change is less clear. I argue that companies in the United States have evolved away from effective teamwork in the product development process, for the same reasons that teamwork has become more important. Although greater number of parts, greater product variation, greater component complexity, greater automation, greater technological complexity, and shorter product lives have increased the importance of teamwork, they have also caused organizations to become larger and have made it more difficult for U.S. organizations to employ effective teamwork in the product design process. It was easy to effectively use teamwork between marketing, engineering, and manufacturing when each of these departments employed only a few people. However, today's marketing, engineering, and manufacturing organizations are much bigger than they were in the early part of this century, as a direct result of the factors described in Chapter 3.

Consider a product's marketing organization. Increased product variation means there are more market segments to evaluate and more models to consider in each segment. Shorter product lives means the definition of these market segments changes faster. Marketing organizations naturally have grown larger to handle this increased complexity.

Engineering organizations have grown even faster and larger than marketing organizations due to the changes that have occurred since the 1920s. A greater number of parts in a product and greater product variation directly increase the number of engineers needed to design a product, because there are physically more unique parts to design. Greater component complexity, greater technological complexity of products, and automation indirectly increase the number of engineers needed to design a specific product because more specialists are needed. Each specialist handles a specific part of the product or addresses a different issue in the product. Often these specialists are part of a supplier's engineering organization. A greater number of these specialists increases the number of personnel needed to design a product since more managers or other personnel are typically needed to plan, schedule, and integrate their efforts.

Manufacturing organizations have also grown. A greater number of parts means more manufacturing steps and thus more manufacturing engineers. Greater product variation means more production control personnel. Greater component complexity, greater technological complexity of products, and automation mean that a manufacturing organization needs more specialists, who often work for suppliers.

Marketing, engineering, and manufacturing organizations have grown to such an extent that the people involved rarely work in the same room. They may not even be in the same city or state or country. One often has to look several levels higher in these functional organizations to find someone who is responsible for integrating the output from these functions. Since important personnel often work for suppliers, integration can rarely depend solely on direct lines of authority.

The result is that it's no longer easy to have effective teamwork between the personnel who need to be involved in the product development process. It's much easier for each functional department or supplier to do its task and then pass it on to the next department; that is what a U.S. firm's organizational structure suggests is the proper mode of operation. Typically, the marketing department first comes up with a new set of product specifications, then the engineering department attempts to develop a product that can meet them. The manufacturing department tries to make the product and finally the sales department tries to sell it. Each of these functions optimizes the product from its own viewpoint; the result, however, is often a product that is not optimized from a system perspective, because optimal design decisions may not have been made with respect to the market's needs and the product's manufacturability.[5]

The Team Approach to Product Design

Most people agree that a more team-oriented approach is needed in new product development. Nevins and Whitney call it "concurrent design."[6] Schonberger calls it a "design partnership."[7] Hayes, Wheelwright, and Clark call it "overlapping problem solving."[8] The Department of Defense calls it "design for producibility," and the Society of Manufacturing Engineers calls it "simultaneous engineering." The common characteristic of these approaches is their emphasis on teamwork between the various functional departments. Suppliers and customers are also often part of the design team. Each department or organization is involved with the project as early as possible. There are phases to the project, but they are not assigned sequentially, each to a different department. Instead, each functional department is involved the entire way through the project.[9]

The Importance of Learning

A number of observers have noted that learning is also an important part of the team-oriented design approach. Lessons learned in one project are applied across projects and to future projects. According to Hayes, Wheelwright, and Clark, "Although many firms can point to one or two particularly successful projects, few achieve superior performance on a consistent basis. Even fewer show steady improvement in their development efforts over time. Yet, getting better and better at product and process development is the key to an enduring competitive advantage."[10]

An essential part of learning is the development and continual improvement of the new product development process. Just as

JIT manufacturing has become an accepted approach to continually improving the production process, firms should continually improve the product development process. Cycle time, a key part of JIT manufacturing, should be an important performance measure of this process. Researchers have found that development time is an important predictor of a product's success;[11] moreover, a strong correlation has been observed between development time and development cost.[12]

The only way to define and improve a product development process is to write it down, measure it, and then improve it. In other words, procedures are needed. But Americans don't like procedures, and many might argue that one of the biggest problems with U.S. firms today is their excessive reliance on procedures. However, the problem may be more in the way U.S. firms develop and use procedures than in the concept of procedures themselves. Procedures are typically developed by management and used by employees. When mistakes happen, management typically adds more complexity to the procedures. For example, Hayes, Wheelwright, and Clark describe what happens when procedures are used in a product design process: "As mistakes are made or unexpected difficulties encountered (for example, someone forgets to double-check something or does not get a necessary sign-off), managers add steps to the procedures to make sure that those problems do not recur. As a result, procedures become more and more cumbersome, and project performance deteriorates."[13]

There is an alternative approach to this development and use of procedures, however. Just as production workers and manufacturing engineers should be made primarily responsible for reducing the cycle time of the production process, so also design engineers, marketing personnel, and manufacturing engineers could be made jointly responsible for reducing the cycle time of the product development process. These employees could write the procedures and use the lessons from each design project to improve these procedures. This obviously requires a lot of teamwork among these employees. They must work together to define and improve a process that often has been done the wrong way — sequentially — for years.

Teamwork for Software Design

A number of writers have argued that software design also needs to become more team oriented and that learning should be an important part of software development. The reason for this is that software programs are becoming more complex. One study found that the number of lines of code in an average business application increased from less than a thousand in 1970 to more than 10,000 in 1980. The same study predicted that the average business application would require more than 90,000 lines of code in 1990.[14] Larger software programs mean that more programmers are typically working on a given business application, and this makes teamwork more important. Further, custom-built software is becoming more popular, which means that programmers must work more closely with customers. Many firms may begin to see software houses in much the same way as they view the suppliers of their fabricated parts and assemblies.[15]

Due to the increased complexity of software programs, a number of software experts have begun to call for a better definition of the software design process. They argue that the software design process must be measured and controlled like any other business process.[16] For example, a study from the Carnegie-Mellon Software Engineering Institute argues the need for a better definition of the software design process, in which there is a repeatable, improving process. It defined five stages in a software design process:

1. initial
2. repeatable
3. defined
4. managed
5. optimized

The study also looked at a number of actual software design processes in U.S. firms and found that most firms are not using very sophisticated processes. Most firms were at the first stage, a small number were at the second stage, a few were at the third stage, and none were at stages four or five.[17]

Some researchers have found that Japanese firms, on the other hand, have developed sophisticated software design processes. Applying the definitions developed at Carnegie-Mellon, the case material presented in Part Five describes a fourth- or fifth-stage software design process. One could conclude from Michael Cusumano's descriptions of the software design processes used by Japan's leading computer makers that these firms have also developed a fourth- or fifth- stage process.[18]

CHAPTER SIX

The Growing Importance of Teamwork in the Technology Development Process

TEAMWORK has also become more important to the performance of the technology development process. This process includes those activities associated with developing and applying new technology to products and processes. It begins with decisions to fund the development of new ideas, methods, or techniques. It ends with the actual use of these new ideas, methods, or techniques in new products and processes. Laboratories, corporate headquarters, suppliers, customers, and the engineering, manufacturing, and marketing departments at individual divisions are involved with the technology development process.

The Changing Nature of Manufacturing Firms

A greater number of parts, greater component complexity, greater technological complexity in a product, and shorter product lives increase the importance of teamwork primarily because they increase the number of personnel involved with the development of specific technologies. A greater number of parts means there are more researchers involved with developing new technology for a specific product. Since greater component complexity means that more specialists are needed to develop the technologies embodied in a specific component, it also increases the number of researchers involved with developing the technologies needed for a specific product. Frequently these specialists are employed by suppliers who produce such products as advanced materials or electronic components. Greater component complexity also increases the importance of the relationships between customers and suppliers because it often increases the number of levels in the supplier-manufacturer-seller chain.

Greater technological complexity means there are more technologies that affect a single product and more products that are affected by a specific technology. For example, automobile manufacturing now depends on various software and hardware technologies associated with factory automation. Yet these technologies can affect virtually any manufacturing industry. Due to this greater technological complexity, the number of researchers concerned with the specific product or technology has increased dramatically, making teamwork more important to the success of the technology development process.

Product lives have also become shorter, which means that things need to happen faster. A greater number of parts, most of which have increased in complexity, and a greater number of technologies must be developed in a shorter time frame. Therefore, more people must not only work together, they must work together more intensely than they did 70 (or even 10) years ago.

These factors have dramatically changed laboratories and the way they have developed new technology over the last 50 to 100 years. Consider America's earliest laboratories. Thomas Edison founded some of this country's first industrial laboratories in the late nineteenth century. Thomas Hughes describes how Edison surrounded himself with highly capable people "whose interests and capabilities complemented his own. Despite their presence, however, it was Edison, as inventor-entrepreneur, who pulled most of the strings of the complex system."[1]

In the early twentieth century, larger laboratories were built by some of the largest corporations in the United States, particularly by the automobile firms. By the mid-1930s, half of America's 200 largest companies had established laboratories. However, a few people still dominated these laboratories; this was possible because there were still a small number of technologies embodied in a product. For example, Stuart Leslie describes the approach of Charles "Boss" Kettering, head of General Motor's research division from 1920 to 1947. According to Leslie, Kettering dominated the laboratory for many years. He used his personal authority to fight administrative wars against production divisions and even had time to be personally involved with technical work.[2]

Today, U.S. laboratories can no longer be dominated by a few individuals. Modern labs are too large and contain too many technical specialties for any one individual to dominate for even a few years, much less for more than 20 years, as Kettering reportedly did. In the 1930s, there were 1600 industrial laboratories in the United States, employing a total of 33,000 people, or an average of 20 people per laboratory.[3] Although there are still laboratories that employ only 20 people, the United States today has many industrial laboratories that employ a hundred times as many people. America's 10 largest firms probably employ more people in their research laboratories than were employed in all of the country's industrial laboratories in the 1930s.[4]

Managing the Organizational Linkages Associated with Technology Development

Laboratories have grown in size in response to the increasing complexity of products, components, and manufacturing processes and to the shorter product life cycles. These factors have increased the importance of teamwork while the increasing size of laboratories and firms makes it more difficult to implement teamwork. One way to look at the importance of teamwork is to consider the linkages that need to exist within a laboratory and between the laboratory and other parts of a manufacturing firm. Consider linkages between

1. a laboratory and corporate headquarters
2. a laboratory and a division
3. specialties within a laboratory
4. firms and their suppliers
5. related businesses

Linkages Between Laboratories and Corporate Headquarters

First, there must be a tight linkage between a corporate laboratory and corporate headquarters. As Roland Schmitt, former vice president for research at General Electric, observes, "The success of corporate-level R&D depends on a thorough understanding of corporate goals and strategies to guide a balance of its programs among various businesses."[5] Unfortunately, this isn't always the case. According to Schmitt, "All too often, there is no agreement between corporate R&D and top corporate management on what the lab's mission and purpose are. In many companies, technical managers seek to promote first-rate science while the corporate office initiates study after study to find out why new businesses have not resulted from centralized R&D."[6] The increasing technological complexity of products and shorter product life cycles have made and will continue to make this linkage difficult to manage.

Linkages Between Laboratories and Divisions

Second, there must be a tight linkage between laboratories and divisions in both the planning and implementation of technology development. According to Schmitt, "You have got to have the [divisions] in the kitchen with you when you're cooking up technology."[7] Schmitt notes that the linkage between the laboratories and [divisions] is not very tight: "Too much of what is done in the U.S. is chuck wagon technology — the technology guys dish it out and say to manufacturing, 'here, do something with it.'"[8]

Dorothy Leonard-Barton and William Kraus make a similar point: "Selling top management on the case for new technology — without simultaneous involvement of user organizations in the decision-making process — is not enough. It is equally important for users of an innovation to develop 'ownership' of the technology."[9] According to Schmitt, "you got to have that partnership on the front end. You have to bang heads with operations until you identify the critical paths of the critical businesses."[10] Unfortunately, this isn't always the case. According to Norm Johnson, vice president of R&D for Weyerhaeuser, "Too often, the U.S. approach is 'let's not waste time, let's get to work.'"[11] These linkages will continue to become more difficult to manage as products include a wider variety of different technologies and product life cycles continue to become shorter.

Linkages Between Specialties Within a Laboratory

Linkages must also exist between different scientific disciplines within a laboratory. After all, synergism between different scientific disciplines is one of the primary reasons for having corporate laboratories. However, the increasing complexity of new products makes it more difficult to achieve this synergy. Today's new products typically depend on a multitude of new technologies. According to Edward Hennessy Jr., Allied-Signal's former chairman and CEO, "To design a new material, you need more knowledge than can be found in any one field of specialization. You need chemists who know what molecular changes are needed, physicists who can measure and understand bulk properties, and engineers who can figure out how to make new materials in a cost-effective way."[12] Even scientific breakthroughs require increasing synergy between different disciplines. Lee Rivers, director of corporate planning at Allied-Signal, states: "Increasingly the real breakthroughs are coming at the crossover points between the sciences."[13]

Linkages Between Firms and Their Suppliers

Suppliers are also an important part of the technology development process. Porter maintains that "competitive advantage emerges from close working relationships between world-class suppliers and the industry. Suppliers help firms perceive new methods and opportunities to apply new technology. The exchange of R&D and joint problem solving lead to faster and more efficient solutions."[14] Studies of specific industries have also come to the same conclusion. In a study of innovations in scientific instruments, Eric von Hippel found that 75 percent of recent innovations have come from users. For example, semiconductor chip manufacturers have accounted for two-thirds of recent advances in the equipment used to make the chips.[15] These improvements obviously require close relationships between customers and suppliers. As the MIT Commission on Industrial Productivity concluded, "It is incumbent on producers not only to try to identify their customers' needs but also to pay close attention to improvements that customers themselves have made or are suggesting."[16]

The growing complexity of components has increased and will continue to increase the importance of teamwork between suppliers and firms. New types of integrated circuits, new materials, and new manufacturing equipment will increasingly be developed by those firms that have close relationships with their customers or with these types of suppliers. These technologies are too expensive and complicated for a firm to develop them independent of their customers. Ted Kumpe and Piet Bolwijn argue that components have become so important technologically that firms need to think about vertical integration more in order to enable and encourage suppliers to make critical investments in new technology.[17]

Linkages Between Related Businesses

Linkages between related businesses is also an important part of the technology development process. Porter maintains that "supporting and related industries" are one of the four most important determinants of a country's competitiveness in a specific industry.[18] Hayes and Wheelwright make a similar observation that appropriate horizontal integration is needed to fully benefit from relationships between a firm and its suppliers. In their example, the electronic component and system maker are in the same firm. They argue that the firm "needs to consider not only the operations and components unique to that product but also the opportunities it has to exploit its experience with other products. Cumulative experience in component distribution, integrated circuit manufacturing, and consumer product distribution are probably all relevant to the new product. This other experience may not only benefit the new product but it, in turn, may add significantly to the cumulative experience in some of these shared tasks, thereby improving the costs of [older products]."[19] In other words, not only is teamwork important between a component supplier and a systems producer, it is also important between related businesses.

The increasing complexity of these individual electrical components will continue to increase the importance of the linkages between these related businesses. Producers of both semiconductors and the types of electrical products that are integrated into large electrical systems will need effective linkages with related businesses to produce products that effectively complement other products in the complex electrical systems.

Is Innovation an Individual or a Group Process?

Many readers may be asking, "Why are these linkages so important? Since research requires creative thinking, isn't it an individually oriented activity? New ideas come from very creative people." I believe that's true: new ideas do come from creative people. But let's look at those new ideas within the context of the technology development process. An important consideration is the difference between invention and innovation. Invention probably requires a great deal of creativity. However, innovation requires a number of people from laboratories and divisions to solve a number of small problems as technology is translated into a working product or process. Countries that are successful at invention are not necessarily successful at innovation.

The Nobel Prize

Consider the Nobel Prize. Citizens of the United States consistently win more Nobel Prizes each year than any other people in the world. However, Americans forget that not so long ago, it was the United Kingdom that invented things and the United States that turned those inventions into new products and processes (i.e., innovation). In fact, until World War II, citizens of the United Kingdom received more Nobel Prizes than citizens of the United States, despite the difference in population size.

When the population differences are factored into this comparison, it has been shown that the United Kingdom still produced more Nobel Prizes per capita than the United States until about 10 years ago. From 1941 to 1960, the United States had 42 Nobel Prize winners, while the United Kingdom produced 16 winners. When adjusted for population, however, the United Kingdom had the equivalent of 56 Nobel Prize winners during this 20-year period. From 1961 to 1980, the United States had 66 winners and Great Britain had 77 when adjusted for population differences. Only since 1981 has the United States passed the United Kingdom in the average number of "population-adjusted" Nobel Prize winners. (Some observers believe that the smaller number of U.K. Nobel Prizes is due to cutbacks in university research funding initiated by Prime Minister Thatcher in the late 1970s.[20])

Despite this record, it has been years since the United Kingdom was considered successful at innovation. Although the United Kingdom is obviously still successful at invention (based on their Nobel Prizes), they are no longer successful at innovation. In terms of per capita income, the United States passed the United Kingdom in the late nineteenth century.[21] Few people have considered British companies adept at innovation in the middle-to-late twentieth century. Somehow, continued strength in invention hasn't helped the United Kingdom remain strong in innovation. Obviously there is a difference between invention and innovation.

So what does innovation require? How does it work? Unfortunately, the technology development process is still not well understood. Psychologists have done some interesting work on how individuals actually innovate.[22] But as for how all of these individual innovations actually fit together, little research has been done. One of the most readable descriptions on this facet of the history of innovation is James Burke's *Connections* (many readers will be more familiar with Burke's public television series by the same name).[23]

Connections

Burke describes how innovations are linked together through a series of somewhat accidental connections. For example, although popular lore portrays James Watt as the sole inventor of the steam engine, Burke observes that Watt merely performed the last stage in its development. The steam engine was the culmination of a long line of developments aimed at pumping water, primarily out of coal mines. Between 1654 and 1705, the vacuum pump, piston pump, steam pump and pumping engine were developed, all by different people. Thomas Newcomen developed the pumping engine in 1705, and Watt's steam engine came from adding a separate condensing unit to Newcomen's engine. Watt's innovation was a critical step, since it meant that one cylinder was not used to both create and condense the steam, which was very inefficient. Nevertheless, it was only one of many innovations that were needed to make a useful steam engine.

Further, Watt's modification of Newcomen's design would not have succeeded without two other innovations that occurred in two completely separate industries:

- Breweries developed the technology necessary to use iron as opposed to brass cylinders in the pumping engine (brass cylinders were too expensive).
- Cannon makers developed a boring machine that could make more precise cannons and cylinders.[24]

Watt applied these innovations to his steam engine, but they were developed by other people. He merely applied their inventions to a new problem.

It's unfortunate that Western history tends to focus on individual development, in this case Watt's steam engine, as opposed to all of the inventions needed to develop a new product. Watt's work is certainly worthy of praise, but it is debatable whether his improvement was more important than all of the developments that preceded and accompanied it. According to Burke and other observers of innovation, the actual connections between each of these innovations are more important than any single innovation.

A natural conclusion from this discussion is that if you want to speed up the development of technology, you need to increase the frequency with which these connections are made. Better methods of communication are certainly part of increasing the speed of these connections. Burke describes how the printing press, mail service, the telephone, and professional societies increased the speed of these connections.[25] Ultimately, however, as the rate of technological innovation increases, monthly, weekly, or even daily communication is required between the key scientific disciplines to accelerate the rate of these connections. This is why laboratories including scientists and engineers from a variety of disciplines were created in the late nineteenth and early twentieth centuries. However, the rate of technological innovation has continued to increase since these laboratories were first created. New methods are needed to integrate the efforts of these research personnel. Today, the rate of technological innovation is becoming more dependent on teamwork.

Communication Networks and Innovation

Other evidence for the importance of teamwork in innovation also exists. One study attempted to discover why some research engineers in a particular laboratory were more productive than other engineers in the same laboratory. The study found that the laboratory's most productive engineers had developed the best communication networks. The best engineers had figured out not only who the experts were in each specialty but also which questions to put to which people. It wasn't just a matter of always asking the top expert in a particular specialty, however, since the engineers would quickly use up their favors that way. These engineers had to save the toughest questions for the top experts and find other people who could answer the easier questions.[26] One lesson from this study might be that engineers need to be outgoing, aggressive, and able to get along with people. Another lesson might be that laboratories that develop good communication networks will most likely be the most productive laboratories. And cultures that encourage this type of teamwork might also end up with highly successful research laboratories.

In summary, many of the activities associated with the technology development process require a great deal of teamwork. Although many of the truly great research breakthroughs (e.g., Nobel Prize-winning research) will likely continue to depend on highly creative individuals, the actual application of new technology to products and processes is becoming more dependent on teamwork. As product lives become shorter, and component and product technology become more complicated, it is becoming increasingly important to focus on the connections between technologies and innovations. At the same time, however, shorter product lives and greater product and process complexity increase the number of personnel associated with the technology development process. This makes it more difficult for firms to have effective linkages within and between laboratories, corporate headquarters, and factories.

Unlike in the production and product design processes, the factors described in Chapter 3 affect the technology development process in both the discrete and nondiscrete parts industries. Although a greater number of parts in a product and greater component complexity are primarily relevant to the discrete parts industries, two of the other factors — a greater technological complexity and shorter product lives — affect both the discrete and nondiscrete parts industries. Therefore, while increases in the importance of teamwork have probably become most pronounced in discrete parts technology development, it has probably also become more important to this process in the nondiscrete parts industries. The increasing importance of teamwork to the nondiscrete parts industries suggests that Japanese firms may become more competitive in these industries in the future. Although Japanese firms may not have an advantage (or at least not a large one) in production and product design for these industries, the increasing importance of teamwork to the technology development process will probably help them become more competitive in these industries in the future.

CHAPTER SEVEN

The Increasing Importance of Teamwork in Three Manufacturing Industries

I T IS INSTRUCTIVE to look at several key industries — automobiles, semiconductors, and electronics — to see how teamwork has become more important to each of these three primary processes.

The Automobile Industry

Ford was the leading automobile maker in the early 1920s. Its design philosophy was to build one basic model year after year with a few add-on options. It produced about one million of these "standard" cars each year during the early 1920s; each car consisted of a few thousand parts. Due to the few product options, there were few feeder lines that were not connected to the main assembly line, and there were very few changeovers. Production workers performed simple assembly operations and had little role in measuring and improving quality.[1] Most parts were produced internally; suppliers primarily provided raw materials such as steel, tires, and glass. The design of Ford's cars and the technology behind it was dominated by a few experts such as Charles Sorenson, P. E. Martin, and C. Harold Wills. Henry Ford himself was closely involved with the design of several of his cars, including the Model A in 1927.[2] These individuals identified themselves more with automobiles than with any specific scientific specialty.

Today Ford is obviously a completely different company. In 1989, not counting Lincoln-Mercury and its other subsidiaries, Ford produced 13 different models, averaging 150,000 cars per model. Each car has over 30,000 parts, many of which are highly complicated castings, moldings, and integrated circuits made by a large number of specialty suppliers. Each type of car may have over 50 different options that can be purchased in many possible combinations.

The automobiles are produced in highly automated factories, with numerous feeder lines to handle the extensive variety of parts and assemblies that are part of one car model. Highly skilled workers operate much of the equipment, and many of them are expected to measure and improve quality. Changeovers are frequent, and many of the production workers are expected to participate. Suppliers produce a large variety of the complicated electrical and mechanical parts, and these suppliers are merely one level in a multilevel system of customers and suppliers.

The development of a new automobile is a highly complex activity; it takes five years to translate a concept into a deliverable product. Needless to say, the president of Ford Motor Company and just a few experts are no longer involved with the design of a particular automobile. It is probably necessary to go at least three levels down in Ford's organization to find someone who is closely involved with automobile design. Further down in the organization, hundreds of engineers are directly involved with the design of a new automobile; if suppliers are included, more than one thousand people may be indirectly involved with the design.[3]

Huge laboratories develop the new product and process technologies needed for automobiles. Since almost every new technology affects an automobile, these laboratories employ thousands of specialists. Electronic controls, new materials, factory automation, mechanics, combustion, software, and sensors are just a few of the technologies that concern carmakers.[4] The technological specialists in these fields often identify themselves more with their scientific specialty than with automobiles.

The automobile industry has changed significantly since the days of Henry Ford. No longer can a few experts handle the product and process complexity associated with a modern automobile. Producing a top-quality car requires teamwork throughout a company's factories, design centers, and laboratories.

Integrated Circuits

Although automobiles were one of the most complicated products being manufactured in the 1920s, they are somewhat simple compared to other products being manufactured in the 1990s. It is safe to say that aircraft, computers, integrated circuits, and other electronic products are more complex than automobiles due to the relative level of maturity in the automobile industry. Integrated circuits are one of the most complex products manufactured today. Further, since they are used in most of today's mechanical and electrical products, they increase the complexity of many other manufactured products.

Integrated circuits are the best example of the recent dramatic increase in component complexity. Integrated circuits did not even exist until the late 1950s; the number of transistors per chip increased by a factor of one million between 1960 and 1990. This increased component complexity has required the development of a very sophisticated production system.

Integrated circuits are produced in more than 200 manufacturing steps, which must be performed in an ultraclean environment. Each of these steps can introduce defects, which means that each step must have a very high yield to produce a high yield of finished packaged chips. A 70 percent chip yield requires each process step to have an average yield of 99.6 percent. Each step requires a sophisticated piece of machinery that contains numerous parts and lines of software code, each based on a different scientific discipline. For example, a wire bonding machine that attaches wires from a chip to its package has more than 10,000 parts and 35,000 lines of software code.[5]

However, a wire bonding machine is based on a very different technology than that used in plasma etching, package molding, or photolithographic equipment. Each of these types of equipment depends on a number of different scientific disciplines. This is a good example of the increase in the number of technologies that affect a single product. Wire bonding equipment depends on ultrasonic vibrations and mechanics. Plasma etching equipment depends on such scientific disciplines as physical chemistry and plasma physics. Package molding depends on technologies related to plastics. Photolithographic equipment primarily depends on optics. Electronics and software are the primary technologies that these types of equipment have in common.

This equipment is linked together into a computer-integrated manufacturing system to improve process control and eliminate human intervention from the factory (people being the main source of dust, and thus defects). Due to the benefits of automated production, semiconductor factories are becoming one of the most complex types of production facilities in existence. Assuming that a wire bonder has a level of complexity similar to that of other semiconductor manufacturing equipment and that five pieces of equipment are needed for each process step, a

CIM system for producing semiconductor chips would require the integration of more than 10 million parts and 35 million lines of software code. Additional lines of software code are needed to integrate the various equipment.

Such a system requires an immense amount of teamwork. No individual specialist or group of specialists can understand such a system. Defects can enter at any of the more than 200 process steps. Moreover, most of the processes are interdependent; a change in one process many affect several other processes, yet each process requires specialists from completely different disciplines. Many of the 10 million parts or 35 million lines of software code can potentially cause equipment malfunction or downtime, both of which can cause yield to suffer. Production workers, process engineers, equipment engineers, and laboratory personnel must work closely together to successfully operate such a system.

The development of new integrated circuits requires even more teamwork, however. Since many new integrated circuits, particularly memory chips, put more transistors on a chip, smaller circuit lines and therefore new processes are needed for each new generation of memory chips. A new memory chip typically requires the development of a new factory such as the one described above. The role of equipment makers cannot be overstated. For a new factory to contain the most recently developed equipment, new equipment must be designed as a new chip is developed. Equipment makers, laboratory personnel, process engineers, and design engineers must work together to design new factories and chips simultaneously. Since the firm that first develops a new memory chip is typically the most successful in terms of yields and profitability, teamwork between these equipment makers, laboratory personnel, process engineers, and design engineers can be the primary determinant of success in the semiconductor industry.[6]

Electronic Products

Another example of the increased importance of teamwork can be found in electronic products. The increasing capability of integrated circuits is giving them a major role in driving performance. These chips are key components in such products as consumer electronics, computers, telecommunications apparatus, instruments, aircraft, automobiles, appliances, and factory automation.[7]

Because chips are key components in these products, the product designers must understand the chips and the chip designers must understand the electronic products. This is particularly true in the case of custom chips such as application-specific integrated circuits (ASICs), which are designed for a specific product. This is an example of the increasing diversity of modern products. As more and more chips are developed for specific applications, the chip designer or company is becoming an important part of any electrical product design team. The success of an electronic product depends increasingly on teamwork between the company that designs the chip and the firm that makes the product.[8]

However, in the same way that semiconductor chips are not directly beneficial to society without their inclusion in an electrical product, many electronic products are also not useful until they are integrated with other electrical products. Consider, for example, the products produced by some of the high-technology industries shown in Table 1-1 and Figure 1-1. Many computers, office products, telecommunications equipment, and instruments are beneficial to society only when they have been integrated with other products produced by the same or other firms. Computer hardware, software, and networks must be efficiently integrated. Customers also often want other office products to be integrated with computers. Further, several people have postulated the integration of computers with consumer

electronics products such as high-density televisions.[9] And producers of photographic equipment are trying to integrate their
products with VCRs. Telecommunications equipment is by its
nature an integration of different equipment, frequently from
different firms. Instruments are increasingly used in integrated
systems such as computer-integrated manufacturing systems,
electrical power plants, and transportation systems.[10]

The increasing importance of integrating these products with
other products makes it more important for the producers to
develop products that complement each other. In other words,
teamwork between these different businesses is becoming more
important, particularly in the design of new products and the
development of new technology for these products. Competitive advantage may come increasingly from the effective integration of these different businesses. Firms that do not have
effective linkages internally or with other businesses may be
leapfrogged by the firms that do.

The design of electronic products is an example of a common
trend in most manufacturing industries in which component
makers are becoming more important members of the product
design team. In the case of electronic products, the increasing
complexity of components can be seen in terms of an integrated
circuit or in terms of the products themselves in that an electronic product is often but one·component in a huge electrical
system. Increasing component complexity means that more specialized employees and firms are needed to design and produce
these components. Greater specialization always increases the
need for teamwork, since more managers are needed to coordinate the activities of these specialists.

Some people have argued that the increased use of application-specific integrated circuits will favor the fragmentation of
the semiconductor and other industries.[11] However, this view
ignores the close supplier-customer relationships that are

becoming more commonplace in the United States. Many U.S. firms are adopting this strategy and almost every management consultant advocates these close relationships. One of the criteria of the Malcolm Baldrige National Quality Award is a close customer-supplier relationship. Such close relationships do not exist in fragmented industries in which companies come and go. Such relationships are based on long-term agreements. The customers and suppliers do not have to be part of the same firm to have these long-term agreements (although it probably helps), but they do have to be willing to make a long-term commitment to cooperation and teamwork in the design of new products and the development of new technology.

CHAPTER EIGHT

Institutional Barriers to Teamwork in the United States

I T IS CLEAR that manufacturing industries have changed greatly since the 1920s. Products and processes have become significantly more complex. Products have a greater number of parts, they have more variations, individual components are more complex, products contain a greater variety of technologies, and product lives are shorter. Each of these factors increases the importance of teamwork in the production, new product development, and technology development processes.

It appears, however, that management philosophy in the United States had, until recently, changed very little. Much of the same system that existed 70 years ago still exists. It is largely based on Frederick Taylor's work in the early twentieth century, which emphasized a rigid vertical hierarchy, economic analysis, and functional specialization. Work is divided into small parts and assigned to specialists. The vertical hierarchy commands and controls these specialists, using a variety of economic tools.[1]

Most U.S. companies have developed organizations based on these rules. Some of this may have been done consciously; some of it may have occurred subconsciously as business schools standardized the nation's management practices. Much of U.S. corporate culture also draws on "rugged individualism" and the myths and folklore that surround American social culture.

Almost every modern discussion of U.S. management holds that Taylor's rules have outlived their usefulness. Tom Peters and Robert Waterman, for example, argued that America's best-run companies are those that least follow these rules.[2] It is difficult, however, for firms to change their corporate cultures. It is argued that the changes that have occurred in these "successful" firms are very small compared to what exists in Japanese firms, since by and large the changes advocated by U.S. writers already exist in Japanese firms. For the most part, vertical hierarchies, economic analysis, and functional specialization still exist in U.S. firms — which they should, to some extent. Even if it is not a perfect generalization, it's useful to look at some of the changes advocated by observers of U.S. management, particularly those changes related to teamwork. Ironically, many of these proposed changes sound very similar to Japan's corporate culture.

Observers have found at least four types of barriers to teamwork in U.S. firms:

1. U.S. firms overemphasize the use of function (i.e., activity) as opposed to output (i.e., product) in their organizations.
2. Vertical linkages are much stronger than horizontal linkages in terms of responsibility and communication.
3. U.S. firms use too narrow a definition of responsibilities.
4. U.S. firms encourage their employees to develop an excessive level of functional specialization.

An Overemphasis on Functional Organizations

Manufacturing firms in the United States generally overemphasize the use of functional organizations. Although firms are often divided into divisions that produce a specific product line, individual divisions largely organize their employees by function. Employees are segregated into functional departments and functional sections, with various staff groups responsible for integrating the various functional activities. Taylor emphasized functional specialization and the use of staff to control these functions. According to Hayes, Wheelwright, and Clark, most U.S. organizations still emphasize specialization and staff groups; they observe that a typical manufacturing firm has three levels of these staff groups in its organization — at the corporate level, the division level, and the factory level.[3]

The overemphasis on functional organization causes problems in the production, new product development, and technology development processes. In the production process, Schonberger argues, U.S. firms should adopt a customer- or product-oriented organization to facilitate the flow of work. He argues that the functional organization encourages employees to form "gangs" and exhibit ganglike behavior. In this situation, "customers at the next process are a rival gang. So are suppliers. So, probably, are the support staffs."[4] Wickham Skinner makes a similar argument that most U.S. factories need to divide themselves, both organizationally and physically, into "plants within a plant and each plant should be dedicated to a specific product or type of product."[5]

The functional organization also causes problems in the new product development and technology development processes. Stalk argues that high-performing Japanese and U.S. firms use a more product-oriented organization to minimize the number of times a development project must cross organizational lines. Gerald Susman and James Dean maintain that by placing product and process designers closer together in the organization, product designs will be more manufacturable.[6] In the technology development process, the functional organization inhibits the flow of technology from laboratories to divisions. For example, Robert Calahan, president of Ingersol Engineers, Inc., argues that firms in the United States should get rid of their existing organizational structures: "Forget the organizational structure we've used for 300 years. Simply put together people who can get the job done, regardless of their function."[7]

A Lack of Horizontal Linkages

A second barrier to teamwork in U.S. firms is that since U.S. organizations emphasize a vertical hierarchy, they often lack horizontal linkages between different functional departments. Taylor emphasized a vertical hierarchy, and for the most part, U.S. firms still seem to think of responsibilities and communication patterns primarily in terms of a vertical organization. According to Hayes, Wheelwright, and Clark, "A worker or a first-line supervisor who wants to communicate with his or her counterpart in another plant, with an engineer in the divisional R&D group, or with a salesperson has no natural channel for doing so."[8]

Horizontal linkages are needed in the production, new product development, and technology development processes. To ensure quality in the production process, Imai notes, "there must be smooth communication among all the people at every production stage."[9] Hall argues for short daily meetings: "Everyone received the same signal from the daily huddle, and it was a good place for little side conversations that prevented the need for other meetings and phone calls."[10]

Hayes, Wheelwright, and Clark, Hall, and Schonberger all maintain that problems need to made more visible. Statistical process control is one way to make problems visible.[11] As Schonberger notes, "A plant using statistical process control is set up for visible management. Anyone can make a tour in an hour or two and know what is right and what is wrong.[12]

Visibility, however, violates a cultural view that employees should be given information only if they have a specific need to know.[13] Information is power, and managers in vertically oriented firms are well aware of this. Information is typically well guarded, and segmentation of responsibilities makes it easier to protect information.

Horizontal linkages are also needed between different functional departments in the new product development process. Due to the emphasis on vertical linkages, U.S. companies often have poor cross-functional communication. Adequate linkages do not exist between different functional departments because most organizations were not designed with these linkages in mind. Yet many observers have found that good communication is a key factor in all successful design projects. Hayes, Wheelwright, and Clark describe the communication aspects of a team-oriented design project: "Development is characterized by extensive overlap, with continual two-way interchange of information at low levels. Different phases of the project are integrated through a shared understanding of the primary purpose(s) of the product or process. Fast, effective problem solving and early conflict resolution are the rule."[14]

Cross-functional communication is also needed in the technology development process, particularly along the five linkages discussed in Chapter 6. Communication is needed within and between laboratories, corporate headquarters, divisions, related businesses, and suppliers. As William Shanklin and John Ryans, Jr., observe, "Face-to-face, in-person interaction and an agenda for meetings are most productive. Marketing and R&D people should talk almost daily during the initial planning effort for a new product or application and regularly thereafter for updating and revision."[15] The linkage between R&D and the marketplace is often the weakest because of geographical separation between R&D and divisions. Most U.S.

research laboratories were built far from factories to free researchers from the pressures of day-to-day factory problems,[16] and communication between laboratories, headquarters, and factories can suffer.[17]

Narrowly Defined Responsibilities

Manufacturing firms in the United States like to have a close match between responsibility and authority, and they like responsibilities to be independent of each other. This is due primarily to the emphasis on vertical organizations, because these vertical organizations like to have responsibilities neatly divided and matched to a particular person. Hayes, Wheelwright, and Clark found that most U.S. firms have developed customized performance measures to evaluate a manager only on the tasks that he or she has the authority to perform. They concluded that these types of performance measures are not capable of encouraging the type of performance needed in today's environment.[18]

Responsibilities need to be broadened and shared in the production, product design, and technology development processes. In the production process, employees and managers alike need to have their responsibilities and performance measures broadened. Some writers refer to these types of employees as multifunctional workers, employees with a broad perspective, or generalists. Hayes, Wheelwright, and Clark call it job enlargement and argue that responsibilities must be shared between downstream and upstream activities.[19] Imai makes a similar argument that "any job involving multiple workers has gray areas that do not belong to any one individual. Such gray areas must be taken care of by whoever is at hand. When the worker sticks to his own job description and refuses to do any more than what is formally required of him, there is little hope of *kaizen* [continuous improvement]."[20]

Managers involved with the production process also need to have their responsibilities broadened. Since responsibility is expected to match authority, most U.S. factories emphasize narrow measures of performance. According to Skinner and Hall, most U.S. firms place too much emphasis on equipment utilization and direct labor efficiency.[21] Both of these factors enable managers to be responsible only for those things they can control, since managers can be made responsible for the equipment and labor in only their part of the factory. To improve system performance, however, these managers must focus on improving the entire production process. Cost, cycle time, and quality are frequently mentioned as the key measures of performance.

Performance Measurement Problems in the Product Design Process

Narrow performance measures also cause problems in the new product development process. These performance measures make it difficult to solve problems that require interaction between people from different parts of an organization. As Donald Heany and William Vinson note, "Our culture has emphasized individual achievement. Measurements provoke behavior patterns that make teamwork difficult and personally unrewarding."[22] Hayes, Wheelwright, and Clark argue that responsibility should be shared between upstream and downstream activities in the product design process.[23]

The problems of measuring the performance of the product design process are very similar to the problems of measuring the performance of the production process. In factories, production decisions often are made to keep direct labor efficiency and equipment utilization high. In a product design process, functional managers often make the same type of decisions. Engineers are moved from project to project to keep everyone working on a project with an acceptable "charge number." The accounting system calculates which functional departments and sections have the highest ratio of productive (worked on a valid charge number) to nonproductive time (not worked on a valid charge number).[24]

Although managers attempt to keep a group of engineers working on the same project, the system of performance measurement encourages them to do otherwise. Engineers enter and leave in the middle of projects despite the problems of getting engineers up to speed. This type of system obviously inhibits the team-oriented approach to product design because it assumes that product design does not really require much teamwork. Yet it is becoming increasingly clear that teamwork should be the focus of today's new product development process.

Performance Measurement Problems in the Technology Development Process

Performance measurement is also a problem in the technology development process. The short-term orientation of most U.S. firms is one aspect of this problem, but this is just the tip of the iceberg. Most laboratories make the same mistakes in measuring the performance measurement of the technology development process as factories make in measuring the production process and functional departments make in measuring the new product development process. In laboratories, functional managers place too much emphasis on labor efficiency, since most laboratories measure a department's or a section's performance by its ability to meet budgets. Departments that bring in more money than they spend are successful, and vice versa. As in the design process, the accounting system calculates which functional departments and sections have the highest ratio of productive to nonproductive time.[25] Therefore, development engineers also are moved from project to project to keep everyone working on a project with an acceptable "charge number." Studies show that this type of excessive rotation has a severe impact on project performance.[26]

While laboratory managers attempt to focus on other issues, such as the linkages within and between the laboratory, corporate headquarters, and the factories, their performance measures tell them to do otherwise. These measures tell them to focus on budgets, not on maximizing the amount of technology that flows from the laboratory to the factories. The performance measures tell them to move people around and play accounting games to make their budgets look good.

These performance measures are a direct result of that part of U.S. corporate culture that believes that responsibility should equal authority. Budgets are emphasized because managers can usually control them — unlike the linkages within and between laboratories, corporate headquarters, and divisions. Such linkages cannot be handled by any one person — they require a great deal of teamwork among a large number of geographically separate organizations. Laboratories could probably learn something from factories on how to measure performance. Like factories, laboratories should focus more on their cycle time (the time it takes to move technology from the laboratory to the marketplace or a factory).

Overspecialization

The concept of specialization is a very important part of U.S. corporate culture. People are typically paid and rewarded according to their level of specialization. Training programs also emphasize specialization, and the emphasis appears to be increasing. The MIT Commission on Industrial Productivity concluded that "functional and disciplinary divisions among technological and managerial professionals have, if anything, become more pronounced." In fact, executives interviewed by the MIT Commission felt that "technical professionals have little understanding of the total production system and often seem more dedicated to advancement within their discipline than to furthering the goals of their firm."[27]

Overspecialization causes problems in the production, new product development, and technology development processes. Schonberger argues that production workers need to be cross-trained so they can handle a greater number of tasks.[28] Hayes, Wheelwright, and Clark state the need for corporate-level training programs that emphasize preplanned job rotation.[29] Hirotaka Takeuchi and Ikujiro Nonaka maintain that the team approach to product design is best suited to "nonexperts" who "are willing to challenge the status quo."[30] Some writers have taken this argument one step further, holding that the U.S. educational system inhibits teamwork in the product design process because engineers are not taught as part of their university education how to work in teams.[31] Therefore, the objectives and rules of the game have to be explained before the team-oriented approach to product design can be used.

Electrical engineers in the United States also seem to believe that the U.S. educational system should place greater emphasis on aspects related to teamwork. In a poll by the Institute of Electrical and Electronic Engineers, a large percentage of respondents claimed that their education did not adequately develop their communications and human-relations skills. About 70 percent of the respondents felt that their electrical engineering educations were inadequate in these two areas, while only 27 percent felt that their technical educations were inadequate.[32]

These four generalizations — an overemphasis on functional organizations, a lack of horizontal linkages, narrowly defined responsibilities, and overspecialization — obviously simplify the changes needed in U.S. management. Many other factors affect U.S. firms and can be used to describe how they operate.

Some firms perform better than others, and some perform differently from others in these four areas. Part Three, however, will describe how Japan's corporate culture encourages Japanese firms to be *substantially* different in these four areas. Parts Four, Five, and Six will describe how corporate culture helps Japanese firms manage the production, product design, and technology development processes.

PART THREE

The Elements of Japan's Corporate Culture that Support Teamwork

(WITH CHRISTINA HAAS
PENNSYLVANIA STATE UNIVERSITY)

We know that Japanese firms have corporate cultures that are different from those found in the United States. Books, articles, anecdotal comments by visitors in trade magazines, and detailed analyses by academics have confirmed the existence of many differences between the social and corporate cultures in the two countries. Although there is disagreement about the details, there is a great deal of agreement with respect to several general differences between Japanese and U.S. corporate cultures.

We also know that U.S. firms must drastically change their organizations to become better at teamwork. As described in Part Two, teamwork is becoming more important to the success of manufacturing industries, and many managers and management researchers are calling for fundamental changes in the way U.S. organizations are managed. Many of these proposed changes are strikingly similar to what Japanese firms are doing and have been doing for many years. Moreover, Japanese firms are doing these things more extensively than most of the managers and researchers expect U.S. firms ever to accomplish.

Part Three summarizes the elements of Japan's corporate culture that support teamwork. Since my examples are drawn primarily from one firm, I do not argue that all Japanese firms have exactly the same type of corporate culture described here. Based on a six-month experience with another Japanese firm, shorter visits to yet other Japanese firms, other studies, books, articles, and conversations with other American visitors to Japan, however, I believe that the differences between Japanese and U.S. firms with respect to these elements of corporate culture are far greater than the differences between and within Japanese firms. I use detailed examples from one firm to show how several basic elements of Japanese culture facilitate teamwork in the three processes outlined in Part Two. These elements of Japan's corporate culture include a product-oriented organization, visible management, oral communication, and extensive training.

Product-oriented Organization Japanese manufacturing firms tend to organize departments and sections by product (or output) rather than function (or activities). This product-oriented organization minimizes the number of times one of the three processes — production, product development, or technology development — crosses sectional or departmental boundaries; thus it enables smaller organizational units (e.g., sections as opposed to departments) to have more control over these processes. Employees tend to be dedicated to a product more than a function. These multifunctional employees work together to develop and improve procedures for performing the production, product development, and development processes. Further, sections or departments that must cooperate in a specific process often share responsibility for the performance of that process.

Visible Management Visible management is used in manufacturing firms and other organizations[1] to present important information about these processes in a simple, standard format. The accessibility of visible management means that everyone, not just those people who management has decided have a "need to know," has access to pertinent information. Examples of visible management exist both in factories and in offices; they include kanban cards, status boards, standardized status forms, and public files.

Oral Communication Within the Japanese manufacturing firm, oral communication patterns facilitate teamwork in the production, product development, and technology development processes. A prime facilitator of oral communication is the open office — large open rooms housing employees of all levels who work in the same department. Since departments and sections are organized by product type and the office layout matches the organization chart, it is easy for employees to communicate readily and frequently with one another; informal conversations continually occur in such an environment. In addition, when two or more sections or departments share responsibilities for production, product development, or technology development processes, a formal communication network exists to support these shared responsibilities. Suppliers, customers (both internal and external), and research laboratories are often part of these formal networks.

Extensive Training Extensive training programs introduce new employees to their department and its highly structured environment. These new employees include recent college graduates and employees transferred from other parts of the firm. Recent graduates learn about the company and the department and, along with transfers, receive extensive training on a department's products, procedures, and technologies. This training enables the members of a department to share a common understanding of these things, even when the actual members of the department change.

CHAPTER NINE

———

The Product-oriented Organization

ONE OF THE most important differences between Japanese and U.S. firms is in the way these firms organize themselves. Companies in the United States tend to organize their departments and sections by functions such as engineering, manufacturing, marketing, or sales. Within these sections, each employee is assigned specific functional responsibilities. Labor efficiency — whether of workers in a factory, engineers in a design department, or researchers in an R&D laboratory — has typically been the key performance measure in a functional organization. Attention to more system-oriented measures, such as total cycle times for the production, product development, and development processes, is often secondary. For these and other reasons, as discussed in the Part Two, many management researchers believe that functional organizations in the United States inhibit teamwork and that U.S. firms should adopt more product-oriented organizations.

It is easier for many Japanese firms to emphasize the system-oriented measures of performance in these processes because they use a product-oriented organization. In this type of organization, a firm organizes its departments, sections, and groups more by product line than by function. This enables these departments, sections, and groups to have control over a larger percentage of the three processes.[1] For example, Table 9-1 compares the number of product-oriented and functional-oriented departments at several Japanese works (factories that also contain design and marketing capabilities). Each of these works contains a significant number of product-oriented departments; a study of a comparable U.S. firm would show very few divisions that contained any product-oriented departments.[2]

Table 9-1. Number of Product-oriented and Functional-oriented Departments at Several Works*

Works	Number of Product-oriented Departments	Number of Functional-oriented Departments
#1 Works	6	3
#2 Works	12	11
#3 Works	7	7
#4 Works	3	5
#5 Works	9	10
#6 Works	4	7

* A works is equivalent to a division in a U.S. firm. They typically contain between 1000 and 3000 employees.

Japanese firms also tend to organize sections and groups by product more than by function. Table 9-2 shows the number of product-oriented and functional-oriented sections and groups within several departments in a Japanese company. As the sample indicates, there are more product-oriented than functional-oriented sections and groups; in a comparable study of a U.S. firm, few departments were found that contained *any* product-oriented sections.

**Table 9-2. Number of Product-oriented and Functional-oriented
Sections and Groups in Several Departments**

Works	Number of Product-oriented Sections and Groups	Number of Function-oriented Sections and Groups
#1 Department	7	2
#2 Department	19	1
#3 Department	4	1
#4 Department	5	1

Clark and Fujimoto also found that Japanese firms tend to organize themselves by product more than U.S. firms, particularly in the product development process. They found that Japanese automobile firms use "heavyweight product managers" and dedicate employees to specific product types more than either U.S. or European automobile firms do. They differentiate heavyweight product managers from both lightweight product managers and functional organizations, in that heavyweight product managers are given a great deal of responsibility and power to integrate the various activities associated with developing a new automobile. They also found that organizations that use these heavyweight product managers tend to organize functions around specific product types (e.g., body engineers around a body type, such as small cars) more than organizations that use lightweight product managers or are purely functional organizations.

Other evidence of a strong Japanese emphasis on product-oriented organization also exists. Many American managers have noticed that Japanese firms have fewer levels of management in their organizations than U.S. firms; the most famous example is the comparison between Nissan and Ford during discussion of a joint venture. A major reason for the fewer levels of management is that Japanese firms have numerous product-specific departments and sections reporting to a single general manager or department manager. For example, according to Table 9-2, one works has 23 different departments reporting to one general manager, and one department has 20 different sections or groups reporting to one department manager.

Contrast this with a U.S. division that uses a strict functional organization. Since there are only a few different standard functions (e.g., manufacturing, engineering, marketing, sales, accounting), it is impossible for a U.S. division that uses a strict functional organization to have more than 6 or 7 (much less 20 or 23) different departments reporting to a single general manager. The emphasis U.S. firms place on functional organizations therefore prevents them from obtaining an organization that is as flat as the organizations used by Japanese firms.

There are three subelements in my definition of the product-oriented organization: multifunctional employees, employee developed procedures and plans, and shared responsibilities.

Multifunctional Employees

The concept of multifunctional employees is a subelement of the product-oriented organization, because multifunctional employees are specialized more by product than function. Many American managers and researchers have noted this difference between U.S. and Japanese employees.[3] Ouchi, for example, argues that Japanese employees have less specialized career paths than U.S. employees; he found that employees in an individual section perform multiple tasks as opposed to each person specializing in one task.[4] As discussed in Part Two, many observers believe that U.S. employees are too specialized.

Employee Developed Procedures and Plans

A second subelement of the product-oriented organization is the use and improvement of employee developed procedures and plans. Multifunctional employees develop detailed procedures for performing processes, and small groups continually work to improve and streamline these employee developed procedures. These employees also create detailed plans for incrementally improving products and the individual technologies contained in these products. The development and improvement of employee developed procedures and plans is a good example of what Ouchi calls "collective responsibility."[5] A group of employees is responsible for developing and improving a set of procedures for performing all or part of a production, product development, or technology development process.

Improving these employee developed procedures and plans is very important to Japanese firms and to Japanese workers. This continuous improvement is what really differentiates the use of procedures in U.S. and Japanese firms. Although U.S. firms often develop procedures, these procedures are rarely created by the employees who will use them.[6] This task typically falls to staff specialists, who rarely obtain much feedback from the users. As a result, the procedures are often ignored or misused; rarely are they improved.

Another way of describing this aspect of employee developed procedures and plans is to say that Japanese firms emphasize processes. Since multifunctional employees develop procedures in order to improve processes (e.g., the three macro-level processes described in Part Two this subelement of the product-oriented organization could be called process improvement or an emphasis on process. A number of observers have argued that Japanese firms emphasize processes more than U.S. firms. As discussed in Part One, Pascale and Athos called it a difference in style; they argued that Japanese firms emphasize the development of problem-solving techniques whereas U.S. firms tend to focus on output. Imai maintains that Japan is a "process-oriented society," whereas the United States is a "results-oriented society."[7]

The MIT Commission on Industrial Productivity also concluded that Japanese firms emphasize process improvement more than U.S. firms. They based their conclusion partly on a study by Edwin Mansfield that found that Japanese firms allocate more of their research and development funds to process improvement than to product improvement.[8] Mansfield found that Japanese firms allocate fully two-thirds of their R&D funds to process improvement, whereas U.S. firms allocate only one-third of their R&D funds to process improvement.[9]

Shared Responsibilities

Production, product development, and technology development processes do of course cross some organizational boundaries within a Japanese manufacturing firm. When this occurs, shared responsibilities (the third subelement of the product-oriented organization) are necessary. These responsibilities, shared between departments and sections within a firm and with external entities, such as laboratories, customers, and suppliers, are explicitly defined. The concept of shared responsibilities is similar to Ouchi's concept of "collective responsibility," in that several people are responsible for a set of activities.

At first glance, the actual sharing of responsibilities appears to be organizationally complicated. However, since the product-oriented organization minimizes the number of times a process must cross different departmental or sectional boundaries, it is possible to explicitly specify each of these relationships and their attendant shared responsibilities. The product-oriented organization and these explicit relationships enable Japanese firms to improve the process as a whole, rather than merely optimizing segments of the process. And as discussed in Part Two, U.S. firms need to develop broader measures of performance and encourage employees to share responsibility for improving them.

Product Orientation in Japanese Factories

It was in Japanese factories that American observers first noticed the use of the product-oriented organization. Abegglen and Stalk found that Japanese firms use "focused" factories more than U.S. firms. These factories are dedicated to producing a specific product line. They employ few support personnel and thus have low overhead costs.[10] Schonberger found that Japanese factories are organized by product flow, whereas U.S. factories have historically been organized by equipment function. Sometimes called "continuous flow manufacturing," organizing a factory by product flow, is one of the central elements of JIT manufacturing.[11]

Multifunctional workers and employee developed procedures and plans are also important parts of the Japanese factory. Multifunctional workers are able to perform most of the operations necessary to make a family of parts in a manufacturing cell. Employees within a particular manufacturing cell are responsible for improving the operation of that cell by working to develop ever-better procedures. Since they are responsible for the quality of the parts produced by their manufacturing cell, they use procedures such as statistical process control to improve quality. Similarly, they are responsible for reducing setup time and making other productivity improvements. They often define and improve the actual operations needed to make the family of parts, and they use line balancing techniques to reduce the amount of waiting time between each worker's operations.[12]

When specific manufacturing cells work together, Japanese factories can closely manage the shared responsibilities because parts require fewer movements between different areas of the factory. Just as responsibilities are shared between interdependent groups, sections, or departments in a Japanese firm, responsibility for reducing inventory and improving quality is shared between different manufacturing cells. Kanban cards are used to manage the amount of inventory held as a buffer between two manufacturing cells, and quality control circles are used to improve quality problems that involve multiple manufacturing cells.

Product Orientation at Mitsubishi Fukuoka Works

The product-oriented organization is also important in Japanese offices. A good example of its use both in factories and in offices can be found at Mitsubishi's Fukuoka Works. As shown in Table 9-3, this works has nine departments. Six departments are each responsible for a single product line, although two of these departments represent different functional responsibilities in the same product line (power integrated circuits and power device production). Each of these six product-line departments is responsible for one or more of the three major processes — production, product development, and technology development.

Table 9-3. Number of Employees per Department at One Works

Department	Number of Employees
Nonproduct	
General affairs	70
Materials	35
Manufacturing control	135
Product	
Power integrated circuits	100
Wafer production	900
Assembly production	450
Power device production	200
Semiconductor equipment	200
Industrial equipment	200
	2300

One of the important benefits of a product-oriented organization is reduced overhead costs. Multifunctional employees can perform more of the activities associated with a specific product; fewer overhead personnel are needed than in a functional organization. Overhead costs represent 80 percent of the nonmaterial costs in U.S. factories; these overhead costs — often due to specialized personnel — represent a smaller percentage of product costs in Japanese firms.[13]

The works outlined in Table 9-3 certainly seems to bear out the generalization that Japanese firms have lower overhead. It has only three functional (nonproduct) departments, which contain fewer than 10 percent of the site's employees. This low number of overhead personnel is particularly striking since many of the functional employees are involved with employee services not ordinarily found in U.S. firms. For example, one-third of the employees in the materials department are responsible for the company store. The other 20 employees are responsible for the site's relationships with its suppliers (e.g., vendor evaluation). Most of the employees in the general affairs department work in the site's health center or organize the intersite employee training programs. Most of the employees in the manufacturing control department are responsible for developing quality standards and for performing various laboratory analyses. Indeed, even the functional workers at the Fukuoka Works are multifunctional.

Employee developed procedures and plans also play an important role at this works. Each of the product-oriented departments has developed a continual improvement program. For instance, the wafer production department continually strives to reduce its cycle time through improved procedures (this is described in more detail in Part Four). Operators are responsible for developing these improved procedures and the engineers work with them on implementing the improvements.

Since many improvements are made through quality circles, operators share responsibility for many of them.

The product-oriented organization is also evident in the product development process. Three of the six product-line departments at this works are responsible for product development and its associated functions. The industrial equipment department, for example, has complete control over both its production and its new product development processes. It contains two engineering sections, a manufacturing section, and a business group. Shared responsibilities and employee developed procedures are crucial for this department's new product development process. According to Kazuhiro Kawabata, previously an electrical engineer in this department, the marketing group works with the engineering and manufacturing sections and with purchasing to concurrently design new products. Each of these sections and groups is part of a multifunctional team that is responsible for designing and developing a low-cost, high-quality product.[14]

Example: The Semiconductor Equipment Department

A second and more detailed example of how the product-oriented organization has an impact on the product development process can be found in the Fukuoka Works' semiconductor equipment department (SED). As shown in Figure 9-1, the SED is divided into three equipment engineering sections, a manufacturing section, a development group, and a business group. Each equipment section's design engineers are multifunctional in that they perform many of the functions traditionally associated with purchasing, marketing, sales, design, and test engineering. Since they deal directly with the department's assembly and parts suppliers, they basically perform the purchasing function. The SED design engineers share responsibility with these suppliers and with the manufacturing section for developing low-cost equipment.

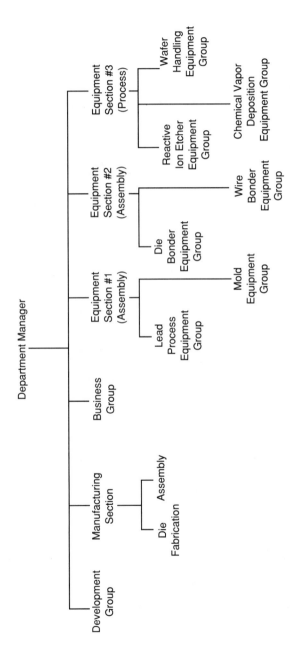

Figure 9-1. Organization of the Semiconductor Equipment Department

A study of Japan's automobile industry also found similar results. Toshihiro Nishiguchi found that Toyota often used engineers as purchasing agents. He found that in 1987 there were only 337 dedicated purchasing agents in Toyota versus 6,000 such employees in General Motors. Although Toyota produces only 40 percent as many vehicles as General Motors, since it purchases many more of its components and materials from suppliers than GM does, it might be expected that Toyota would have as many purchasing agents as General Motors.[15]

The design engineers in the semiconductor equipment department also perform many of the activities traditionally associated with sales and marketing functions. Since the department has a very close relationship with its major customer, Mitsubishi's semiconductor chip business, there are no marketing or sales personnel in the department. The multifunctional design engineers interact directly with the semiconductor process engineers; they basically share responsibility with these engineers for satisfying a specific customer order, designing new equipment, or defining the needed equipment technology. The small business group merely handles the paperwork associated with these transactions.

Employee developed procedures and plans are also important in the SED's product development process. The department's design engineers have together developed a number of procedures for performing this process. They have developed highly detailed procedures for designing, manufacturing, and testing a new equipment model; three working groups within the department continually update and improve these procedures, as well as introduce new tools to the design process. The purpose of these procedures and the working groups set up to improve them is to reduce design cycle time, improve equipment quality, and facilitate the standardization of software modules, electrical hardware, and mechanical technologies.

Effect of the Product-oriented Organization on Individual Responsibilities and Activities

The product-oriented organization evaluates employees and managers differently than does a functional organization. Similarly, employees and managers in a product-oriented organization perform different activities than do those in a functional organization. This section presents an extended example that illustrates some of the differences between product-oriented and functional-oriented organizations by examining the criteria for evaluation and the activities of managers and employees of the semiconductor equipment department.

Shared responsibilities have a strong effect on how managers in this department are evaluated. Responsibility for equipment design and the development of new equipment technology is shared between the department and several other organizations within Mitsubishi. These shared responsibilities tend to increase the number of criteria by which the SED is evaluated. These criteria include equipment performance patent applications and patents acquired, in addition to profits.[16]

Profits are not the dominant criteria, primarily because they do not represent the department's full contribution to Mitsubishi's semiconductor chip business. This attitude is best expressed by one of the department's group leaders: "Profits aren't that important because the semiconductor equipment department is so small in comparison to Mitsubishi's semiconductor business. And if there are any profits, they should go to the equipment users, since the semiconductor equipment department really only exists to serve the semiconductor business."[17]

Put another way, Mitsubishi's total profits depend more on how fast the SED can help the semiconductor factories install new and appropriate equipment than on the department's actual profits. Mitsubishi's factories buy most of their equipment from the SED: this long-term commitment and relationship enables them to have substantial input into the department's equipment design and technology development decisions. In this way, the factories obtain the equipment that best suits their needs and schedules.

The specific criteria by which the SED is evaluated depends on the perspective of the group performing the evaluation. The headquarters of Mitsubishi's semiconductor business is concerned with how rapidly the SED responds to the needs of the semiconductor business, while the semiconductor factories evaluate the department by how well its equipment performs for them in their factories. The general manager of the works, on the other hand, measures the SED's performance by how fast its sales are growing.[18]

Effect on Managers

The shared responsibilities and the multiple criteria under which the semiconductor equipment department is evaluated affect the activities of its department manager, Mr. Banjo, and its section managers. Rather than focusing exclusively on the SED's day-to-day operations, Banjo devotes a great deal of effort to the department's external relationships. According to Banjo, he visits laboratories and the headquarters of Mitsubishi's semiconductor chip business about once a week to forecast equipment needs and the number of engineers needed to design this equipment and to determine how best to reduce the cost of future equipment. He makes it a priority to introduce his engineers to engineers at the laboratories with whom he has associated in his 30-year career with Mitsubishi. This is not so much an attempt to help his engineers "get ahead" as it is a way to facilitate the communication and cooperation between his department and the corporate laboratories. In addition, he works to persuade the laboratory personnel to use some of their corporate funds to develop technology that can then be used in semiconductor equipment.[19]

When Banjo is physically at the SED office, he spends most of his time as a facilitator. Although he is minimally involved with individual equipment design projects (usually only in terms of budgets), he is primarily involved with defining the technological future of the semiconductor equipment and the processes used to design and manufacture the equipment. He summarizes for his employees the technological views of the key people associated with Mitsubishi's semiconductor business. With the section managers, he develops long-term equipment plans.

Banjo also spends a great deal of time simply walking around the office speaking with employees. He does not have an enclosed office. His desk is located at one end of the department's open office, which makes him extremely accessible. According to a senior member of the SED, the department manager had in-depth individual conversations with over 70 percent of his engineers (there are about 70) in his first year as department manager.[20]

Section managers within the SED are evaluated in a manner similar to the way the department manager is evaluated. Each section's equipment performance, profits, and number of patent applications and patents acquired are used to evaluate the section manager's performance. However, the section managers are also evaluated according to their contribution to those departmental activities whose objective is to improve the department's performance as a whole. These activities involve the standardization of mechanical technologies, the integration of technology development plans, and the development of improved design, quality control, and production procedures.

The activities of the section managers reflect the multiple criteria by which they are evaluated. According to Mr. Nakamura, the section manager responsible for the wire bonder equipment group, he primarily participates in departmental improvement projects and deals with external organizations such as suppliers, laboratories, and customers. He participates in department meetings and in working groups that standardize mechanical technologies, integrate technology plans, and develop improved procedures and systems. He meets with a president of one of his suppliers about once a month to discuss cost reduction strategies. He visits Mitsubishi's two most automated semiconductor production facilities monthly to participate in productivity improvement meetings. He also meets with section managers from other semiconductor factories to negotiate equipment prices.[21]

Nakamura spends very little time managing the section's day-to-day operations; he rarely attends the weekly equipment meetings or any of the smaller meetings that involve only one equipment group. He stays in touch with his section's activities primarily by attending his section's daily standup meetings (described later), by reviewing various documents such as schedules and equipment orders, and by taking advantage of the open layout of the office, which enables him to hear almost everything that goes on.[22]

Effect on Individual Engineers

The activities and responsibilities of individual engineers are also affected by the product-oriented organization. Individual engineers in the SED also share responsibilities, albeit in a complicated, hierarchical fashion primarily based on seniority. Each engineer shares responsibility for developing, delivering, and improving the equipment in which his group holds responsibility. The product-oriented organization facilitates these shared responsibilities or group identity; it is much easier for an employee to identify with a product as opposed to a function.

For example, consider the hierarchy shown in Figure 9-2, which describes the relationships between the various engineers in one equipment group within the SED, the wire bonder equipment group. There are 11 engineers in this group, including a group leader, Hiroshi Honda (about 43 years old when interviewed) and an assistant group leader, Kazuhiro Kawabata (about 37 years old). Honda is very involved with the day-to-day operations within the group, although he is evaluated by many of the same criteria by which Nakamura, his section manager, is evaluated. Honda and Kawabata primarily represent the group in its regular meetings with external entities such as laboratories, suppliers, and customers.

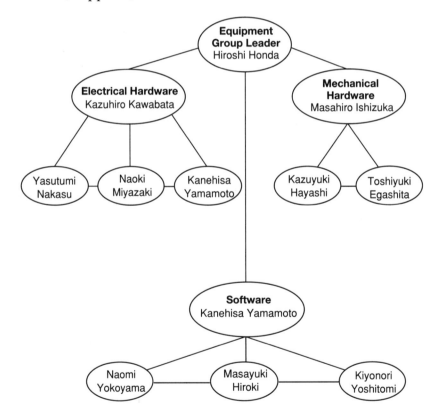

Figure 9-2. Hierarchy of Responsibilities in the Wire Bonder Equipment Group

Three engineers within the wire bonder equipment group share responsibility for developing and integrating the electrical, mechanical, and software designs for each equipment design: Kawabata, Masahiro Ishizuka (about 33 years old), and Kanehisa Yamamoto (about 30 years old). In addition, each of these engineers will typically have the lead responsibility for specific equipment orders. In other words, these engineers also perform the duties typically handled by a program manager in a U.S. firm.

The other engineers, who range from 24 to 30 years old, support these more senior engineers in the manner shown in Figure 9-2. The electrical engineers share responsibility for the electrical aspects of the wire bonder; the mechanical and software engineers operate in a similar manner. In addition, responsibilities are often shared between these electrical, mechanical, and software subgroups. For example, the wire bonder's new bonding head required specific electrical and software efforts that two of the young engineers (Masayuki Hiroki and Yasutumi Nakasu) were familiar with. Therefore, these two engineers from different subgroups shared responsibility for the final development of this bonding head with very little guidance from the senior members of the group.

The product-oriented organization facilitates the integration of electrical, mechanical, and software technologies in the wire bonder equipment group. Each of the employees needed to develop and integrate these technologies is in the group; they sit in adjacent desks in the department's open office. Therefore, it is easy for the wire bonder equipment group leader to allocate these engineers to various tasks on a daily basis. Further, although each of the engineers in the group tends to specialize in one aspect of the wire bonder, each of them knows a great deal about the wire bonder in general. They have worked with the equipment for several years, so they tend to specialize in a product, not in a function.

CHAPTER TEN

Visible Management

THE SECOND aspect of Japan's corporate culture that supports the production, product development, and technology development processes is the management of information. Here, again, there are important differences between U.S. and Japanese firms. I use the term *visible management* to refer to a set of techniques used by Japanese firms to present key information about these processes in a simple, straightforward manner. Rather than determining who has a "need to know" and then disseminating information to those individuals, the techniques of visible management tend to keep all employees in the information loop and make a great deal of information available to almost everyone. Although the phrase *information revolution* may have originated in the United States, American management could learn a great deal from Japanese organizations about how to adequately manage information.

Ouchi's concept of collective responsibility sounds similar to this concept of visible management. Since information is readily accessible, everyone in a group has some responsibility to use and act on the information in a productive manner. Visible management is also a form of implicit control, another aspect of Japan's corporate culture described by Ouchi and cited in Part One. By making key performance measures "visible," everyone has some control, albeit indirect, over the performance.

American observers of Japanese manufacturing have documented examples of visible management in Japanese factories. For instance, the kanban cards used to signal the need for a part or an assembly to be delivered or produced are visible to anyone walking through a factory. These cards have replaced process sheets and other less visible — and less responsive — means of disseminating information. Similarly, many Japanese factories, particularly automobile factories, use buzzers to signal that a product has been completed. Actual production (number of products) in a specific day versus expected production at a specific time may also be displayed prominently in the factory to focus the employees' attention on the purpose of the production process: to produce and ship fully operational units. Another example is the lights that are often used to indicate the status of automatic equipment. Since Japanese workers often operate multiple machines (one semiconductor assembly section at the Fukuoka Works had five workers operating more than 50 wire bonding machines), they need a quick and easy way to spot a machine that needs attention.[1]

Status Boards

Beyond the factory, where their presence has been well documented, techniques of visible management are also used in Japanese offices to display pertinent information concerning design processes, and in the manufacturing area for development projects. Status boards are a pervasive form of visible management in these contexts. Some of these boards are used to indicate each employee's current whereabouts; others display important information about current machine status; still others give specifics on project activities and related problems.[2]

Whatever their specific function, most of these white magnetic boards contain a certain set of fixed categories (which cannot be erased) as well as lines or cells where various types of current information can be added by employees, using pen or magnetic markers, when the status of an employee, machine, or project changes. These boards are displayed in prominent locations where they are readily visible to everyone. I have observed these status boards in over 100 Japanese factories and offices. Their use extends beyond the manufacturing world: Status boards are often used at post offices, medical centers, banks, and other places where employees must be kept abreast of changing information.

Employee Status Boards

Employee status boards give each worker a place to signal whether he or she is in the factory or the office that day and his or her specific location, internally or externally. When employees are needed to share specific information with others, to attend a meeting, or to take a call from a customer, the status board allows coworkers to find them quickly. Since section and department managers tend to be quite busy, special status boards present their monthly schedules, making it easy for employees to plan activities around the schedules of their managers and to keep abreast of important meetings.

At many of the meetings I attended at Mitsubishi and at Yokogawa Electric, information from an individual not present would be necessary for resolving a certain issue. In these cases, we checked the status board, located the individual, received the information, and continued the meeting. The status boards facilitate the interactions that are critical to getting things done. They reduce cycle time and alleviate the frustration of "telephone tag," which slows productivity and decision making.

Project Status Boards

Project information and performance measurements are also summarized on status boards in factories and offices. The wafer, assembly, and semiconductor equipment departments at Mitsubishi Fukuoka Works use these boards to list the source and quantity of each department's major orders. Figure 10-1 shows a translation of one such board that summarizes all equipment orders in the SED. Employees write in the order number and name, quantity ordered, source of the order, delivery date, and weekly work schedule. Magnets bearing the names of standard production activities (e.g., assembly, test, debug) or of the people performing those activities are used to help describe the weekly schedules for each equipment order.

Order Number	Order Name	Quantity	Source of Order	Delivery Date	Weekly Schedule June July August September
	(Employees enter information)				(Employees enter activities and names using magnets)

Figure 10-1. Example of a Status Board Showing Activities and People Responsible for Existing Orders

Equipment Status Boards

The SED also uses two types of status boards to display pertinent information about each equipment order; both are located next to the site where the equipment is assembled, tested, and debugged. One type of board shows the person in charge of the order, the quantity ordered, the source of the order, and any special information about the order. The other board shows the weekly schedule, results, and problems for the same piece of equipment (see, for example, Figure 10-2).

Date	Plan	Results	Problems
4/25	Debug feeder	Completed for Type 1 chips	
4/26	"	Completed for Type 2 chips	Equipment jam (end of lot)
4/27	Debug vision		
4/28	"		
4/29	"		

Figure 10-2. Example of a Status Board Showing the Schedule of an Advanced Packaging System

Performance and Strategy Boards

Performance measures were posted in almost every Japanese office and factory that I have visited. In the SED office, the number of patent applications and patents acquired, and the number and percentage of drawings produced using computer-aided design (CAD) were charted monthly for each equipment group. Further, department and section strategies were also posted and updated as they were changed.

These four types of status boards encourage employees to focus on their department's macro-level issues and processes. The SED's status boards focus on all three major processes, depending on the type of project described: a regular order (the production process), a new equipment model (the product development process), or an experimental device (the technology development process). The status boards also make it easy for everyone to understand the schedules and problems with these projects and make it easy for employees to identify potential bottlenecks in completing the projects. Employees and managers can spend less time distributing reports and schedules and more time solving problems and reducing cycle times.

Status Forms

Status forms, a second form of visible management, are also used to summarize information about production, product development, and technology development processes in a simple but effective way. The SED's design methodology (detailed in Part Five) uses standardized forms to develop the equipment's design documentation. Many of the steps in this process require special forms to be completed; these forms represent a standardized set of documents for each product. This standardization makes the forms easy to read and thus facilitates the reuse of software and other aspects of the design in a subsequent model or product.

Other types of status forms are used to monitor each equipment group's orders. These status forms are often similar to one of the status boards described above. For example, the SED uses a form similar to the project status board shown in Figure 10-1 to describe the status and schedule of all the projects concerning a particular type of equipment. This particular type of form is then distributed at weekly meetings and used as a basis of discussion. Only the macro-level activities (such as mechanical, electrical, and software design, part fabrication, assembly, test, and debug) are shown on these schedules, while more detailed activities are shown on each engineer's individual schedules.

Employees also use status forms to summarize their schedules, both for their own use and for their coworkers. For example, Figure 10-3 shows a form that engineers in the SED use to describe the status and schedule of their individual projects. Based on the projects they are involved with and the macro-level schedules for these projects (e.g., the equipment orders described above), they list their two main projects, the degree of progress in these projects, other work, the reports they are writing, and their daily activities over the month (only two weeks are shown in this example). These individual schedules are discussed in each equipment group's weekly meetings, as the group leader, and to a lesser extent the other senior members of the group, prioritize activities and attempt to identify potential schedule delays. These status forms probably enable engineers to understand each other's schedules and identify potential bottlenecks in meeting the individual project schedules better than engineers who do not use such forms.

Project Name	Degree of Progress*	Week #1 4/1 - 4/8	Week #2 4/9 - 4/16	Remaining Schedule
Feeder Software	Original (Hr) Plan: (from/to)			
	Revised (Hr) Plan: (from/to)			
Vision Software	Original (Hr) Plan: (from/to)			
	Revised (Hr) Plan: (from/to)			
Other Work				
Report Subjects				

* Engineers are expected to fill in the number of hours and dates in which the project runs.

Figure 10-3. A Monthly Work Schedule for an Individual Engineer

Special status forms are used to describe the status and schedule of a project that requires close cooperation between several people in a short time period. These forms typically complement the employee developed procedures in that the procedures define the tasks listed on the status forms. Figures 10-4 and 10-5 present an example of the way in which status forms complement employee developed procedures. These figures explain a detailed labor-hour projection and schedule for eight software problems that involve three engineers in the SED's wire bonder equipment group. Each problem is divided into five activities that are standard tasks in the department's design methodology:

1. define external specifications
2. define internal specifications
3. define hierarchical control process (HCP) charts
4. write software code
5. debug

The combination of this type of status form and the detailed procedure enable the wire bonder equipment group to easily allocate tasks among the available personnel.

Software Problem	External Specs	Internal Specs	HCP Chart	Write Code, Debug
Feeder	#1 Eng. - 1 hr.	#2 Eng. - 2 hrs.	#2 Eng. (2)	#2 Eng. - 2 hrs.
Vision	#1 Eng. (1), #2 Eng. (2)	#3 Eng. - 2 hrs.	#3 Eng. (2)	#3 Eng. - 2 hrs.
Torch Control	#3 Eng. - 3 hrs.	#3 Eng. - 3 hrs.	#3 Eng. (3)	#3 Eng. - 3 hrs.
CPU	#1 Eng. - 2 hrs.	#1 Eng. - 2 hrs.	#1 Eng. (2)	#1 Eng. - 2 hrs.

Figure 10-4. Labor-Hours per Person for a Specific Software Problem

S/W Problem	4/17-4/21	4/24-4/28	5/1-5/5	5/8-5/12	5/19-5/27
Feeder	External Specs	Internal Specs	HCP Charts · Write Code & Debug		
Vision		External Specs	Internal Specs	HCP Charts · Write Code & Debug	
Torch Control			External Specs	Internal Specs · HCP Charts	Write Code & Debug
CPU				External Specs · Internal Specs	HCP Charts · Write Code & Debug

Figure 10-5. Schedule for Solving Software Problems

Public Files

Visible management is also supported by ready accessibility of written information. In most Japanese offices — including government and commercial offices — written or printed information is stored in bookcases that line the outer walls of the office. These cases, or public files, contain files or loose-leaf binders that are labeled according to an internal classification scheme. I have seen these public files in more than 100 offices.

One reason for public storage of information is to save space, since real estate is often as precious in the office or factory as it is in Japanese society at large. Another critical reason for public storage of information, however, is to facilitate information exchange between people working on the same or similar projects. This means that typically only one copy of a document is stored and people access it as the need arises; this contrasts with the typical practice in the United States of copying and distributing documents for individuals to store in their own work areas — whether or not they need to refer to the documents again. In the Japanese firm, everyone has access to information, but the responsibility for organizing and maintaining it is shared across a number of workers.

In the semiconductor equipment department, information ranging from accounting and purchasing data to design documentation to technology development plans is stored in public files. Since engineers in most Japanese product-line departments such as the SED perform a large number of nonmanufacturing functions, each of the equipment groups in the SED develops and maintains its own files for a number of purposes, including accounting, purchasing, and official correspondence. Much of the information in files consists of standardized status forms; forms and files have corresponding numbers for easy access and correct refiling.

Some of the most critical and widely accessed information in the public files consists of design documentation and technology development plans. Since regular employees themselves work together to develop the procedures used to perform their department's and section's processes, it is important that all employees involved (which is indeed almost everyone) have access to similar information and also use similar procedures to enter information in the files. This procedure includes a numbering system that classifies files by type of equipment, model, and other factors.

These files help the department integrate the work from various engineers. The integration of each engineer's design contribution within the equipment groups depends on these public files, since they enable engineers to easily locate other engineers' contributions and define their own. Similarly, the files make it easier to integrate the work of several equipment groups, since the plans and employee developed procedures for each group are available to everyone. For similar reasons, the integration of each equipment group's technology development plans depend on these public files. As will be discussed in more detail in Parts Five and Six, these public files are a critical component of the SED's strategies for standardizing software, electrical hardware, and mechanical technologies; the files make it easy for employees to find documentation and other information for previously developed software, hardware, and mechanical technologies.

The primary benefit of the public files is that they make it easy for employees to find information — information that may have been developed by other employees or other groups but is necessary to perform and improve the production, product development, and technology development processes. Employees don't have to ask other employees for the information and seldom have to spend a great deal of time trying to find it. Since the files are organized by a classification system that everyone understands, the burden for developing and following classification schemes is shared among the entire group. Many of the numerous questions that I asked during my research at Yokogawa Electric and Mitsubishi were answered after someone referred to these public files. When employees didn't known the complete answer or wanted to show me an example, they knew exactly where to find the information because they understood the department's filing system.

CHAPTER ELEVEN

Oral Communication

THERE IS MUCH evidence to suggest that Japanese firms have achieved levels of communication far superior to those of comparable U.S. firms. Most American visitors to Japanese firms are amazed at the levels of communication that exist there. As discussed in Part Two, many observers believe that U.S. firms need to improve their communication, particularly cross-functional communication.

There is also quantitative evidence to support the generalization that Japanese firms communicate better than U.S. firms. Clark and Fujimoto found that Japanese automobile firms have higher levels of internal and external integration than U.S. or European firms. Internal integration refers to how well the firm coordinates different employees in the firm and external integration refers to how well the firm interacts with the market places. Clark and Fujimoto also found that the levels of internal integration correlated with development time and development cost and the levels of external integration correlated with product quality.[1] In other words, the high levels of internal and external integration have a great deal to do with the performance advantages Japanese firms have in the product development process.

Communication is obviously an important part of internal and external integration. Chapter 10 described several nonverbal communication techniques that enhance internal and external integration by making information visible to many employees. This Chapter focuses on oral communication, another important element of internal and external integration.

This element of Japan's corporate culture is somewhat similar to Ouchi's concepts of collective decision making and implicit control. Open offices and formal communication networks, two elements of oral communication, enable more employees to be involved with a decision (i.e., collective decision making) and more employees to have control, albeit indirect control, over a department's performance (i.e., implicit control). The formal communication networks explicitly define who is involved with making a collective decision and who has implicit control over specific decisions. In addition, informal communication networks, a third element of oral communication, support the concept of collective decision making, in that support for decisions is often gathered through these networks.

Office Location and Layout

Japanese factories, offices, and laboratories are located geographically and organized spatially to facilitate communication between parts of the organization that are involved with the same product or process. Since factories are organized by product flow, employees working on subsequent manufacturing processes in a specific product's flow are working adjacent to each other and it is easy for them to communicate. Because Japanese firms favor a product-oriented organization, all of the functions related to a single process are typically located at the same works. Engineering and manufacturing sections are located close to each other to facilitate concurrent engineering of products.

Laboratories are typically located near the works that use their technologies. For example, Mitsubishi's large-scale integrated (LSI), optoelectronics, and application-specific integrated circuits (ASICs) laboratories are all located at the same site as the headquarters for Mitsubishi's semiconductor chip business. Mitsubishi's information system laboratory is located adjacent to its computer works.

Customers and suppliers are also located near one another when possible. For instance, Mitsubishi's semiconductor equipment department, a supplier of semiconductor equipment to Mitsubishi's semiconductor factories, is located at the same site as several of Mitsubishi's semiconductor factories. Mitsubishi's telecommunications business, a customer of Mitsubishi's semiconductor business, is located at the same site as the headquarters of Mitsubishi's semiconductor chip business.

The office seating arrangement for a department within a Japanese firm matches the organization. For example, as shown in Figure 11-1, the SED office is arranged according to its organizational chart. Employees who are part of each of the three engineering sections occupy adjacent seats in the same row. Similarly, those in other groups (development and marketing) also sit near one another. Desks are typically separated only by a 12-inch-high divider, or by no divider at all. I saw this type of office layout in the more than 100 Japanese offices I visited. Many Americans who have worked in Japanese offices — I have spoken with more than 50 — describe a similar situation.

Daily Standup Meetings

One benefit of this spatial organization is that it enables each section or group within the SED to hold meetings while workers remain at their desks. These short standup meetings occur

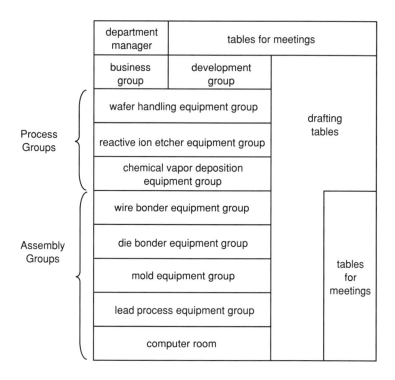

Figure 11-1. Office Layout of the Semiconductor Equipment Department

almost every day, last less than 15 minutes, and take place following lunch. When the bell rings to signal the end of lunch, employees stand at their desks and discuss their section's and group's latest activities in a fairly formal manner. Although the content, form, and length of these meetings differ among Japanese businesses, they are very common in Japanese firms. These meetings existed in almost every Japanese office I've visited; one study estimates that 80 percent of Japanese offices hold these types of meetings.[2]

The SED holds these meetings in a fairly formal way. First, the section manager summarizes the status of the latest equipment installations, the content of past and future meetings in which he has participated, and areas that need improvement. Second, engineers summarize the patents that have been submitted, the content of the working group meetings in which they have recently participated, the purpose and activities of any business trips, and upcoming social events.

Each section then breaks into equipment groups that discuss in more detail those activities specifically related to the equipment group. The equipment group leader discusses the same type of issues raised by the section manager, although he does this only for his equipment type and in more detail. Typically, he summarizes the daily performance of newly installed equipment, such as yield or type of equipment stoppages. The engineers bring up issues they feel are important, such as the outcome of business trips, the status of an important equipment project, or an upcoming social event.

Finally, the chair of the equipment group meeting summarizes his particular job. The chairmanship of these equipment group meetings is rotated so each member participates. Some of this information is repetitive; for example, important meetings, projects, or social events may be mentioned several days in a row. However, there is almost always new information about the performance of newly installed equipment or equipment under development.

Over a four-month period (about 100 meetings), I categorized each topic discussed in the section's and the wire bonder equipment group's daily standup meetings. These data are shown in Table 11-1. Since daily section meetings were not held until November (before that they were held once a week), there is more data for group meetings than for section meetings.

In total, there were about eight topics per section meeting and two topics per group meeting. Topics in the "future meetings" category were typically mentioned in several meetings before they actually occurred. These meetings typically involve the group or section, but sometimes they contain information about the activities of the department or even the general manager. Topics in the "business trips" category are also repeated in several meetings. Every business trip involving a member of the section is mentioned so that each employee of the section knows what other section members are doing. Topics in the problem/status of project category were also repeated several times, although each repetition almost always contained new information.

Table 11-1. Issues Discussed in Daily Standup Meetings by Category

Type of Issue	Section Meeting[1]	Group Meeting[2]
Problem/Status of Project	35	47
Past Meeting	28	24
Future Meeting	59	33
Requests	10	16
Business Trip	6	29
Other	23	31
Total	161	180

1. Data taken from 22 meetings   2. Data taken from 80 meetings

The daily standup meetings are a powerful yet simple technique for keeping everyone informed of important events. Everyone learns of the goals, problems, and strategies of their own group and of the section and department at large. Each employee is given a chance to summarize his own work on a regular basis and coworkers are kept abreast of an individual's progress and problems. Employees hear a summary of almost every meeting in which even one of their members has participated and they typically receive these summaries within a couple days of the event. Even the general manager's activities are often discussed. Social events — of which there are many — are announced and workers are encouraged to participate. In the United States, much of this kind of information could be learned only through the grapevine, leading to frequent misunderstandings and lost information.

Informal Communication

The open office and its layout also facilitate informal communication. Since the employees are not separated by walls or cubicles, it is easy for employees to ask questions of one another and to overhear other conversations. Because the layout of the office matches the product-oriented organization chart, employees are seated next to other employees with whom they should be talking, since they work with them.

Japanese employees spend a great deal of time working together. Table 11-2 summarizes communication pattern data collected within the wire bonder equipment group (which consisted of 12 engineers) at 15-minute intervals over two one-week periods. It shows the percentage of time spent in the office or factory, percentage of office time working alone or with others, and percentage of time with others that occurred in formal meetings or informal conversations. (More detailed information about data collection and results are presented in Appendix B.)

As Table 11-2 shows, engineers tended to spend three-quarters of their time in the office, and of that office time, less than half was spent working alone. Fully 52 percent of the engineers' time was spent working with others, with over three-quarters of that time spent in informal conversations and one-quarter in scheduled meetings. The senior members of the group, as might be expected, spent less time working alone. Mr.

Table 11-2. Communication Patterns in the Wire Bonder Group

	All Engineers	Senior Members
Time spent in factory	24%	27%
Time spent in office	76	73
Office time spent alone	48	24
Office time spent with others	52	76
Time with others in meetings	23	35
Time with others in conversations	77	65

Nakamura (section manager), Mr. Honda (group leader), and Mr. Kawabata (assistant group leader) spent about the same amount of time in the office as did the group as a whole, but only 24 percent of that time was spent working alone. They also spent more of their time with others in meetings (35 percent).

This table does not capture the full dynamics of a Japanese office, however, since it tells us little about the actual interactions of employees in informal conversations. Employees frequently ask each other questions and then return to their work. If the question causes the employees to identify something that is unclear, it evolves into a short impromptu meeting. Frequently, other employees hear something in the conversation that is relevant to their work and they also enter the conversation. As the conversations evolve and move to other related topics, particular members may drop out. These short conversations enable employees to quickly exchange the information needed to ensure that they are all moving in the same direction.

Table 11-3 attempts to capture these informal interactions and more fully characterize the dynamics of a Japanese office. It shows data gathered over two full-day periods for nine members of the wire bonder equipment group. Informal communication patterns were monitored at five-minute intervals, 96 intervals per eight-hour day. Only the time spent in the office was available for analysis; time spent in formal meetings was excluded. Therefore, the mean number of hours in which the data were collected is also shown. (More information about the collection and analysis of the data is presented in Appendix B.)

Americans who have worked for Japanese firms in Japan are typically very impressed by the dynamics of a Japanese office. Watching Japanese employees easily exchange information is a powerful example of teamwork, particularly since meetings often occur at the drop of a hat: Someone asks a question that no one can answer, and everyone decides that a meeting is in order. Because of the open office, the informal conversations, these impromptu meetings — and even the noise level — many American observers get the impression that there is very little structure in Japanese firms.[3]

Table 11-3. Informal Interactions in the Wire Bonder Group Over a Two-Day Period

Mean hours in office	5.0
Number of conversations per engineer, per hour	4.7
Number of separate individuals conversed with per engineer, per hour	2.1
Number of interactions per hour (may include multiple individuals)	6.4
Percentage of intervals engaged in informal interactions per engineer, per day	60%

However, the Japanese office only *appears* to be unstructured. The product-oriented organization and its subelements (particularly employee developed procedures and shared responsibilities) obviously entail a great deal of structure, much of it formalized. There is also a great deal of structure to the communication that occurs in a Japanese firm: Standardized forms and regular agendas for meetings are but two examples. Despite how it might appear to a casual, outside observer, there is probably even more structure, particularly horizontal structure, in Japanese firms than there is in U.S. firms. The open office and informal conversations merely provide flexibility in this highly structured environment.

Formal Communication Networks

Formal communication networks are an example of horizontal structure in Japanese firms in that their purpose is to provide horizontal communication — between different functions and products and with customers, suppliers, and laboratories. These networks exist to support responsibilities that are shared between multiple employees, groups, sections, or departments. Most decisions in a Japanese firm are made using these formal communication networks.[4] Americans often characterize this as "consensus decision making"; Ouchi called it "collective decision making."[5] These terms do apply to the way these formal communication networks operate. However, not everyone is involved with these "collective decisions." Due to the product-oriented organization, each group or section has only a finite number of groups or sections with which interdependencies exist. Japanese firms define the specific relationships in which responsibilities are to be shared between different groups or sections within a firm or between firms; they develop formal communication networks to match those shared responsibilities. Only one or two representatives from each group or section are typically involved with the formal network. Further, these networks consider only those issues in which the two organizations actually share responsibility.

Example: The Semiconductor Equipment Department

Consider the formal communication networks that exist in the semiconductor equipment department. Each equipment group holds weekly equipment meetings, and there are a number of intergroup meetings that support the responsibilities shared between each group in the department.

Weekly Equipment Meetings The wire bonder equipment group's weekly equipment meetings are attended by all members of the group and one member of the manufacturing section. The meetings follow a fairly standard format. The group leader first summarizes what Japanese call *renraku jiko*. This literally means "contact items," and it includes those activities and meetings that are related to the group. Second, the senior members of the group summarize the group's major equipment orders, development projects, and cost reduction activities. Third, each engineer summarizes his individual projects. As described in the last section, status forms are used to summarize both the project and individual schedules. Any problems that surface during these discussions are discussed in more detail. However, in this meeting, the group leader is principally concerned with schedules and staffing needs so problems are primarily discussed in terms of these issues. Typically, smaller meetings are arranged at this time to actually solve the problems.

Intergroup Meetings The SED also has a number of intergroup meetings that support the responsibilities shared between the different equipment and functional groups. These meetings include a bimonthly department meeting for the managers, group leaders, and senior members of each group (about 20 people); a weekly department meeting for section managers; a weekly order status meeting; and bimonthly meetings for each working group in the department. The weekly department meetings are the main avenue for the department manager to communicate his vision and strategies to the department. He discusses these in more detail at biannual department meetings.

The order status meetings review the status of each equipment order; they are attended by each section manager, equipment group leader, and representatives from the marketing group.

The working group meetings are more oriented toward problem solving. While the other departmental meetings primarily discuss broad issues and are basically for information transfer purposes, the working group meetings focus on more specific projects such as improved procedures and plans for the development of the department's core technologies. One working group was implementing an improved order management system, a second group improving the flow of information in the department, and a third group defining the vision technology needs that are common to each equipment type. Each working group includes one member from each of the seven equipment groups in the department.

The organization of the other working groups changed throughout my year in the semiconductor equipment department. These changes reflected the SED's evolving view of the needed improvements in the department. At the end of 1989, there were three working groups: mechanical, electrical, and software; the software group included three subgroups. These working groups typically included one member from each equipment group and they usually met twice a month for a couple of hours. Their purpose was to improve the department's design process and develop methods to facilitate the greater use of standard software, hardware, and mechanical technologies.

In addition, numerous quality circles help these working groups implement improvements in the design process. Although many Americans are more familiar with the use of quality circles in factories, most Japanese offices also use them to improve processes and procedures. In the case of the SED, engineers who are not members of a working group are members of a quality circle. These quality circles meet about twice a month; they primarily implement improvements in the design process by developing better status forms to document the equipment's design.

Example: The Semiconductor Chip Business

Regular meetings are also used to support the responsibilities shared between different departments within Mitsubishi. The semiconductor equipment department is involved with several such relationships, and regular meetings exist to support these shared responsibilities. The meetings most important to the SED are the regular meetings that support the responsibilities it shares with several departments within Mitsubishi's semiconductor chip business.

Responsibility for introducing new equipment into Mitsubishi's semiconductor factories is shared among the semiconductor equipment department, Mitsubishi's semiconductor factories, and the assembly and wafer process engineering departments (organizationally part of Mitsubishi's semiconductor chip business). Five types of communication meetings are held monthly to support these shared responsibilities: equipment; assembly equipment; assembly technology; wafer process equipment; and wafer process technology. The attendees of these meetings are diagrammed in Figure 11-2, and the meetings are briefly described below.

Equipment Communication Meetings These meetings include two design engineers from the appropriate equipment group in the SED, one manufacturing engineer from each of Mitsubishi's semiconductor factories, and a few manufacturing engineers from the appropriate section in the assembly or wafer process engineering department. These engineers discuss the installation of existing equipment and the design of new equipment and their associated schedules. To reduce equipment costs, these equipment communication meetings also attempt to standardize customer needs such as material handling methods.

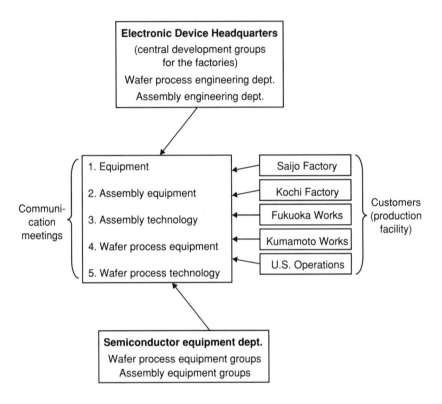

Figure 11-2. Organizations Represented at Communication Meetings

Assembly Equipment and Assembly Technology Communication Meetings These meetings include some of the same people who attend the equipment communication meetings although the topics are a little different. Both the assembly equipment communication and assembly technology communication meetings include one representative from each of the four assembly equipment groups in the SED, two members from each factory, and four members from the assembly engineering department.

The assembly equipment communication meetings discuss several topics also covered in the equipment communication meetings: equipment design, installation, and their associated schedules. The focus of the assembly equipment communication meetings, however, is on the integration of the four types of assembly equipment. The assembly technology communication meeting focuses on the development of new technology for future generations of assembly equipment. New control, vision, or frame handling technology is frequently discussed. Chapter 23 discusses these activities in more detail.

Wafer Process Equipment Communication Meetings Wafer process equipment communication meetings include one representative from each of the three process equipment groups in the SED, two members from each factory, and four members from the wafer process engineering department. The purpose of the meetings is very similar to the assembly equipment meetings but its focus is the integration of the various types of wafer fabrication equipment.

Wafer Technology Communication Meetings The purpose of the wafer process technology communication meetings is similar to that of the assembly technology communication meetings: the development of new technology for future generations of process equipment. However, since process equipment tends to drive the development of new memory chips, the process technology communication meetings focus primarily on developing the process technology necessary for new generations of memory chips. These chips are developed in Mitsubishi's LSI Laboratory. The SED's latest process equipment, such as etching or chemical vapor deposition equipment, is installed first in the LSI laboratory. Therefore, the process technology communication meetings include representatives from the laboratory, the SED, the wafer process engineering department, and other laboratories within Mitsubishi. As discussed in Part Two the faster new process equipment can be developed, the sooner new generations of memory chips can be brought to the market. This formal communication network and the shared responsibilities that this network supports helps Mitsubishi quickly develop new memory chips. Chapter 23 discusses these activities in more detail.

Other Types of Communication Networks in Mitsubishi's Semiconductor Business Departments within the semiconductor business also work closely with suppliers and laboratories. For example, a formal communication network exists to support responsibilities shared between the SED and its suppliers. A formal network also supports responsibilities shared between the department and several laboratories.

The Effect of Formal Communication Networks on the Speed of Decision Making

A number of observers have argued that Japanese firms make decisions very slowly and that their inability to make fast decisions is a major disadvantage for them. Most of these observations are based on business dealings between Japanese and U.S. firms. Management of the U.S. firms is often shocked at how long it takes Japanese firms to make decisions concerning a business proposition such as a joint venture between the two firms.

The reason for this lack of speed in decision making reflects a difference in the way Japanese and U.S. firms view relationships between different organizations. Japanese firms believe in long-term relationships in which concepts of formal communication networks and shared responsibilities are important aspects of doing business. When it is considering a new relationship with an external organization, the Japanese firm believes that a formal communication network and a definition of shared responsibilities are both needed. The development of these networks and the definition of these new responsibilities, however, typically affect the intricate balance of communication networks and responsibilities that already exist in the firm. Therefore, it can take a Japanese firm a great deal of time to modify these formal networks and definitions of shared responsibilities before it can make a decision.

Most internal decisions, however, do not require a Japanese firm to modify its formal communication networks and the shared responsibilities that these networks are set up to support. In fact, since these networks are instituted in the first place to facilitate the necessary decisions, Japanese firms can probably make most internal decisions faster than U.S. firms. These formal networks enable various organizations to stay in close communication with each other. Through these networks, long-term plans are discussed, developed, and continually modified.

Informal Communication Networks

Informal communication networks within Japanese firms complement the formal communication networks previously discussed. Several elements of Japan's corporate culture facilitate the development of informal networks and make them probably more extensive than a similar network in a typical U.S. firm. The open office makes it easy for employees from different groups and sections to interact. At Mitsubishi, job rotation (discussed in Chapter 12), lifetime employment, and the corporate R&D selection process (a type of employee developed procedure) also help develop an informal network of employees within different works.

The informal communication network provides the flexibility needed to complement the rigid structure found in the formal communication network. Within Mitsubishi's semiconductor equipment department, employees constantly discuss intersectional issues such as standardization, the department's design methodology, or the integration of different assembly equipment into an advanced packaging system. Many of these conversations must occur before a consensus can be reached in the regular meetings. Employees with new ideas frequently must convince the key members of the group before the formal meeting in order to have their ideas accepted by the entire group. These informal conversations also facilitate the transfer of technology between different equipment groups within the SED. Part Five describes how this network helped the department apply software used in one type of equipment to another type of equipment.

Corporate-wide networks, however, are the most interesting aspect of the informal communication networks found in Japanese firms. These networks are developed through the continual rotation of employees between related works and laboratories. For example, the SED's department manager has worked in several different organizations within Mitsubishi's semiconductor chip business. As mentioned earlier, he visits one of these organizations about once a week to discuss trends in the semiconductor industry and the semiconductor chip business's strategies to deal with these trends. He claims that one of his primary responsibilities is to introduce his employees to key laboratory personnel in order to bring new ideas into the department.[6] Part Five discusses an example of how this network helped the SED identify the technology necessary for a new image interface system in the die bonder's vision system. These networks are also often the source of new business for a Japanese firm. Part Six discusses how an informal network of employees helped Mitsubishi identify the necessary resources to develop its first facsimile machine.

CHAPTER TWELVE

Extensive Training

IT IS WELL known that Japanese firms provide extensive training to their employees. Both Mitsubishi and Yokogawa provide their employees, particularly new hires, with a great deal of training. This training is used to introduce employees to the immense amount of structure, particularly horizontal structure, that exists in Japanese firms. Without these training programs, it might take employees several years and a great deal of effort to understand this structure. Although training is primarily important for introducing recent college graduates into the firm, it is also used to introduce employees to a new department or new managers to the structure that exists at the upper levels of the company. In addition, cross-functional job rotation, which many Japanese employees undergo at some point in their career, is a form of training in that it exposes Japanese employees to multiple parts of a firm.

Other observers of Japanese firms have also concluded that Japanese firms provide their employees with more training than U.S. employees receive. The MIT Commission on

Industrial Productivity came to this conclusion. Ouchi maintains that Japanese firms emphasize training more than U.S. firms, particularly lifelong training in his concept of "non-specialized career paths." Pascale and Athos make a similar argument when they differentiate between the skills of Japanese and U.S. employees.[1]

Let's look at the example of Mitsubishi's Fukuoka Works, which has more than 30 separate training programs (see Figure 12-1). It has special programs for freshmen (i.e., new employees), technicians, managers, engineers, and the business staff. The subjects range from quality control, English, and office automation to electronics, semiconductors, and mechatronics. The length of these programs ranges from one week to nine months. About one-third of these programs (see asterisks in figure) are conducted at another factory or laboratory.

The training programs shown in the first four columns in the left side of Figure 12-1 are advancement training. These programs are typically one to two weeks in length.[2] Technicians have about five levels of nonmanagerial grades and a few levels of managerial grades. Each grade has its own training program; to move up a grade, employees must successfully complete the appropriate training program.

There are also training programs for section, department, and general managers. These training programs introduce new managers to the roles they are expected to play in their new position. This managerial training also typically includes technical training that will be necessary as the manager moves to a new position.

Functional training, shown in the next three columns, may be closer to what Americans think of as typical training programs. Technicians, engineers, and white-collar employees are taught new technical skills in semiconductors, electronics, mechatronics, and other subjects. The programs for technicians and engineers are typically five to seven months long; the programs for white collar employees are often much shorter.[3] Technicians or engineers often receive this longer training when they are transferred or promoted.

The Semiconductor Equipment Department's Freshman Training Program

As with most Japanese firms, Mitsubishi's freshman training program is the company's most extensive training program; Yokogawa has a very similar program. The Mitsubishi program introduces recent college graduates to Mitsubishi's and their new department's organization, products, technologies, and culture. An accelerated form of this program is also used when employees are transferred into a new department.

The semiconductor equipment department's freshman training program is shown in Table 12-1. In their first month, new employees learn about Mitsubishi as a whole through various lectures at corporate headquarters. In the second month, they learn about the SED's employee developed procedures (e.g., drafting system, design standards), equipment, and technologies (e.g., microprocessors, CAD, motors, materials, C language, painting, heat processing) in an informal classroom setting. Regular employees provide the training; typically several people are qualified to serve as trainers for each class. Since the public files are often used to explain the various equipment, these training programs require very little preparation.

These training programs are particularly important to the SED's design methodology and its efforts to use common technology in each equipment type. As mentioned earlier, the department uses a very formal design procedure. The freshmen learn about this procedure in the discussions about CAD, C language, the drafting system, and design standards. As discussed in Part Five, standardization is a very important part of the

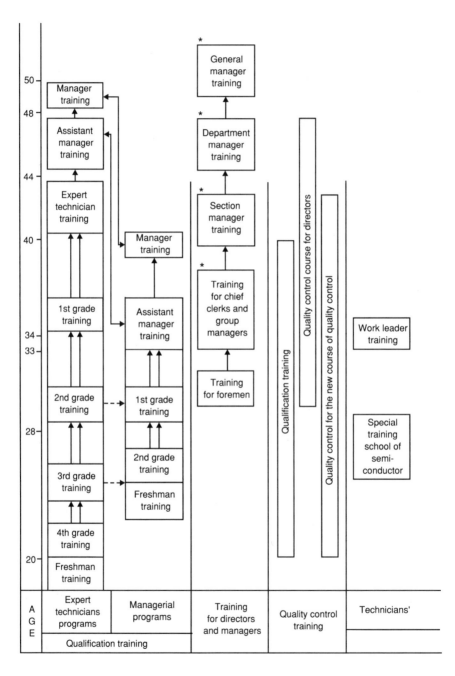

Figure 12-1. Training Programs (Fukuoka Works)

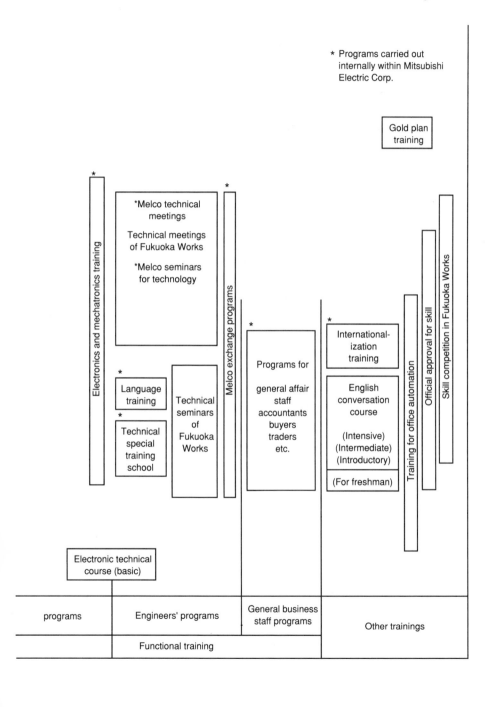

Table 12-1. The Semiconductor Equipment Department's Freshman Training Program

Date	Morning		Afternoon		
April	Headquarters Training				
5/8	Orientation for new appointees		Department rules		
5/9					
5/10	Reactive ion etcher	8085 assembly language	Drafting system	CAD class	CAD practice
5/11	Chem. vapor depos.				
5/12	Wafer handling				
5/13	Wet etcher		Design standards	Models	
5/15	Die bonder				
5/16	Wire bonder	8086 assembly language			
5/17	Mold equip.				
5/18	Dies and molds				
5/19	Lead process	Materials	Factory tour		
5/22	Small motors	Heat processing	Individual study		
5/23	Sensors	C Language	Painting		
5/24			Plating		
5/25-	Individual study				
5/31					
June	Hands-on training in equipment manufacturing				
July	Hands-on training in wafer and assembly production				
June to December	Special project				

department's strategies. The lectures on microprocessors, motors, materials, and the equipment promote this standardization.

In the third and fourth months, the freshmen receive on-the-job training in the equipment manufacturing section and in the wafer and assembly manufacturing departments. The third-month curriculum teaches about the manufacturing processes used to make the equipment and thus promotes *design for manufacturing:* the ability to design products that are easy to manufacture. In the fourth month, the freshmen learn about the department's customers, an obviously important prerequisite to designing new semiconductor equipment.

In the fifth through ninth months, the freshmen are given a small project that culminates in a presentation to the other freshmen and the corporate headquarters training personnel. The preparation for their presentations is a very good example of shared responsibilities, an important element of the product-oriented organization. A freshman's new group feels partly responsible for the individual freshman's performance during the presentations. If an individual does not perform well, employees outside the group will place some of the blame on the group and its members. Therefore, each member of the group tries to help the new employee as much as possible.[4]

Mentors for New Employees

When new employees enter the department, they are also assigned mentors who are expected to help the freshmen through their early careers at Mitsubishi. The mentor is expected to teach the freshmen about technical and nontechnical aspects of the company. Typically, employees who have had good mentors become good mentors. For example, one senior member of the SED felt that he had an excellent mentor when he first entered Mitsubishi and that his mentor had a very important influence on his career. This engineer therefore spends a great deal of time helping the younger employees for whom he is responsible.[5]

Transferred employees also go through some of this training (usually the second-month segment), although in an accelerated and abbreviated manner. I received some of this training when I first began working in the semiconductor equipment department. Mr. Hirayama, who was transferred to the SED from Mitsubishi's computer works, also received this training, some of it at the same time as myself. This particular employee was transferred to the SED to help the department better integrate its equipment controllers with Mitsubishi's semiconductor factory control systems. His first project was to develop a standard equipment controller. Since this project required him to be familiar with the various types of equipment and technology that are made and used by the department, his training paid off very fast.

The purpose of these training programs, particularly the freshman training program, seems to be more than just a transfer of technological information, however. More important, it appears to be a method of introducing a new member to the group's other members. The new member learns who the experts are in the department and can later go to these people for information. Through these people, they are introduced to other people in the department. It probably does not take very long before new employees have met and learned the expertise of most of the department's members.

The other technical training programs also seem to have this purpose. The training programs are almost always internal to Mitsubishi, and the trainers are often employees of the research laboratories. Therefore, the employee also learns who in Mitsubishi has expertise in a particular area and can ask these people for information at a later date. Knowing who to ask for information can lead to quick solutions to problems or long-term projects.

Cross-functional Job Rotation

Job rotation is also a form of training. Japanese firms frequently rotate employees through various parts of the corporation and through various functional positions to develop more well-rounded employees, improve communication, and transfer technology between different parts of the organization. Sometimes the employees return to their original position after a few weeks; sometimes they never resume their previous job.

Employees are transferred frequently between different departments within Mitsubishi's semiconductor chip business. According to Mr. Honda, the leader of the wire bonder equipment group, the main purpose of these rotations is to transfer semiconductor process technology between different semiconductor factories. As soon as a new process technology is introduced into one factory, job rotation is often used to move this technology into Mitsubishi's other semiconductor factories.[6]

Job rotation is also used to improve communication between Mitsubishi's laboratories and factories. According to Dr. Kishida, the former general manager of Mitsubishi's central laboratory, 50 percent of the department managers in Mitsubishi's laboratories have had experience as a manager in a factory. He remembers working in one laboratory where every department manager but himself had factory experience.[7] Job rotation between factories and laboratories enables these organizations to better understand each other's needs and capabilities.

Engineers are also rotated between laboratories and factories. According to Dr. Kishida, after five to ten years of work in a Mitsubishi laboratory, engineers are considered for possible transfer to a factory. In 1985, about 30 percent of these engineers were transferred, although the number decreased by 1989 to between 20 and 30 percent. These personnel transfers occur especially when a technology is being transferred from a laboratory to a factory. For example, engineers were transferred from laboratories to factories when production began on the 4M DRAM and the optical disk. After a few years, about half of these engineers returned to the same laboratory.[8]

Other Japanese firms also emphasize job rotation. Ouchi and other observers have noted that Japanese firms rotate employees between different functional departments more often than U.S. firms do.[9] Kazuo Koike observes that Japanese employees have a series of closely related jobs throughout their careers.[10] Hiroshi Takeuchi notes that Japanese firms rotate new workers through a variety of jobs.[11] Schonberger and Abbeglen and Stalk also make similar observations.[12]

Leonard Lynn, Henry Piehler, and Paul Zahray have done probably the most detailed comparison of U.S. and Japanese engineering career paths. They found that 62 percent of Japanese engineers were rotated at least once in their careers as compared to 35 percent of American engineers. Thirty-five percent of Japanese engineers were assigned at some point to production, compared with only 14 percent of American engineers. And 50 percent of Japanese engineers have served in one outside assignment in research, design, or development activities as compared with only 14 percent of American engineers.[13]

CHAPTER THIRTEEN

Interdependencies Between the Elements of Japan's Corporate Culture

PART THREE has described several elements of Japan's corporate culture that support teamwork. Although not all Japanese firms necessarily have the type of corporate culture described here, I believe that the differences between Japanese and U.S. firms with respect to these elements of corporate culture are far greater than the differences among and within Japanese firms. To understand these issues better, it would be useful to have quantitative measures and data on multiple U.S. and Japanese firms with respect to these elements of corporate culture; further research is needed in these areas.

To assess the relative importance of these elements, we also need comparisons between these quantitative measures and the performance of the organizations. It is important to recognize, however, that the elements are highly interdependent. They tend to reinforce each other and in some cases one element could not exist without another element. In other words, it may be difficult to distinguish the effects of each individual element on an organization's performance.

The interdependencies are illustrated in Figure 13-1. The shaded boxes summarize the main elements of Japan's corporate culture described in Part Three. The other boxes summarize the effects of each element (vertical axis) on the other elements (horizontal axis).

The *product-oriented organization* enables smaller organizational units (departments or sections) to take responsibility for a significant percentage of the three macro-level processes. The third and fourth subelements of the product-oriented organization have a particularly strong impact on the other elements of Japan's corporate culture. Employee procedures depend on status forms. And status forms, in the guise of standard forms and schedules, facilitate the development and implementation of these procedures. These status forms are used to prompt employees with the proper activities for performing a procedure and also to schedule these activities.

Shared responsibilities also require formal communication networks and job rotation. The formal communication network complements the shared responsibilities. Regular meetings, which are part of this network, enable organizations to stay constantly in touch with each other. Without this network, it would be very difficult for different organizations to share responsibilities.

Visible management is used to present important information about these processes in a simple standard format. The subelements of visible management also improve other elements of Japan's corporate culture. Status forms and other types of status boards improve the content of oral communication because these subelements provide a great deal of information about schedules and problems with meeting schedules. Employees do not have to constantly provide the same oral updates to different people. Instead, conversations can focus on solving the problems. The employee status boards, which can be

Effect of: \ Effect on:	Product-oriented Organization	Visible Management	Communication	Training
Product-oriented Organization	1. Product flow 2. Multifunctional employees 3. Procedures 4. Shared responsibilities	1. Open office, public files, and status boards to exist 2. Procedures require status forms	1. Shared responsibilities require formal communication network	1. Shared responsibilities require job rotation
Visible Management	1. Status forms improve procedures	1. Kanban cards 2. Status boards 3. Status forms 4. Public files	1. Employee status boards locate people 2. Status boards and public files improve quality of questions.	1. Public files facilitate training
Communication	1. Formal communication supports shared responsibilities	1. Open office enables public files and status boards to exist	1. Open office 2. Lunch meetings 3. Formal & informal communication network	1. Employees receive continual training through continual communication
Training	1. Job rotation facilitates shared responsibilities 2. Procedures taught in training program	1. Status boards and public files taught in training program	1. Introductions from training programs 2. Mentor program facilitates questions 3. Job rotation facilitates communication	1. Corporate organization, products, and procedures 2. Mentor program 3. Job rotation

Figure 13-1. Interdependencies Between the Elements of Japan's Corporate Culture

easily seen due to the open office, also facilitate these informal conversations because they make it easy to find people.

The public files, another aspect of visible management, facilitate extensive training. Since the public files contain all of the design documentation, procedures and other pertinent information about the department, it is easy to gather the materials for a short training class. Employees do not need to ask other employees for the necessary training materials. They merely look in the files, which are organized for easy access by all employees.

Oral communication patterns, including office layout and location, formal communication networks, and informal communication networks, support the three macro-level processes. Several subelements are also necessary prerequisites for the other elements of Japan's corporate culture. The open office is the key driver of informal conversations, the daily standup meetings, and the mentor program. The open office makes it easy for employees to ask questions and see if specific employees are available. Since a new employee typically sits next to his mentor in the open office, it is easy for the new employee to ask questions of the mentor. The open office also enables the public files to be seen easily by all of the employees, both when they are shelved in their proper place and when they are in use on an employee's desk. Since the employees who work together are seated next to each other in the open office, the public files can be located next to the employees who use the files.

Extensive training programs are used to orient employees who are new to a company or department with their highly structured environments. These programs, particularly the training program for new employees, introduces many of the other elements of Japan's corporate culture. New employees learn about an organization's employee developed procedures, shared responsibilities, status forms, public files, and formal communication networks.

Job rotation is a form of interdepartmental training. It transfers technology in its broadest sense — products, processes, culture — between different departments and enables different organizations to better understand each other. Through this understanding, it facilitates shared responsibilities and increases the quantity and quality of communication between organizations.

PART FOUR

Teamwork in the Factory: Reducing Cycle Time and Implementing Automation

Japan's manufacturing success is well known. Japanese firms carry less inventory, have shorter cycle times, achieve higher yields, and use more automation than U.S. firms. We also know a great deal about how they achieve these goals. Continuous flow manufacturing, short setup times, small lot sizes, pull systems, mixed-model production, empowered workers, statistical process control, design of experiments, and close relationships with suppliers and customers are some of the techniques Japanese firms use to reduce inventory, cycle times, and numbers of defects.

What we don't know are the little yet important things. For example:

- How do specific improvements occur and how do these improvements occur on a continual basis?
- How do Japanese firms implement large numbers of suggestions per worker each year while many U.S. firms struggle even to implement a suggestion system?
- How do operators work together and with engineers?
- How do workers, engineers, and management work with customers and suppliers on continual improvements in cycle time, yields, and cost?

Japan's corporate culture helps Japanese firms do these things. It facilitates continual improvements in cycle time, yield, and cost by promoting teamwork between workers, engineers, and management within an individual firm and between a firm and its suppliers and customers. Examples from one of Mitsubishi's semiconductor factories are used to demonstrate how Japan's corporate culture facilitates continual improvement through teamwork. Some of the examples are drawn from an unpublished report written by Warren Smith, a former Fulbright scholar who studied a semiconductor factory (where I was also located) in Japan.

This is not to say, however, that all Japanese factories are similar to the examples described in this part. There are many differences between Japanese factories; some of these differences are firm-specific while others are dependent on the type of industry that the factory is in. For example, due to the high cost of achieving the ultraclean atmosphere necessary for semiconductor production, chip factories tend to be organized according to clean-room requirements. Photolithographic equipment is kept in much cleaner rooms than diffusion, etching, or assembly equipment, which tends to limit the application of continual flow production. Until more localized means of maintaining an ultraclean atmosphere are developed (e.g., within the equipment), it will be difficult to apply continual flow manufacturing to the semiconductor industry to the extent it is used in other industries.

I will not attempt here to provide a complete description of how Japan's corporate culture facilitates continual improvements in Mitsubishi's semiconductor factories. Given the number of techniques used by Japanese firms to continually improve cycle time, costs, and quality and the way in which Japan's corporate culture affects these techniques, an entire book would be required to adequately address these issues for a case study of even one firm. I will touch only briefly on how Mitsubishi improves its semiconductor yields, although yield improvement is obviously an important aspect of Japan's success in the semiconductor industry. In any case, my objective here is to demonstrate that the same elements of corporate culture that help Japanese firms implement high-performance product development or technology development processes also help them implement high-performance production processes. Although multifirm studies are ultimately needed to determine which elements of Japan's corporate culture most affect the performance of these processes, the elements discussed in this part appear to facilitate significant process improvements.

CHAPTER FOURTEEN

The Role of Japan's Corporate Culture

ALL OF THE elements of Japan's corporate culture described in Part Four help Mitsubishi to continually improve cycle times and quality and to implement automation.

Product-oriented Organization

The first element, the product-oriented organization, enables small organizational units to be responsible for most of the functions needed to fabricate wafers and assemble chips. The wafer production department at Mitsubishi's Fukuoka Works performs all of the operations necessary to fabricate and test a wafer. The assembly production department performs all of the operations necessary to assemble chips into individual packages. (For convenience, I will refer to these as "wafer department" and "assembly department.") Section managers and group leaders in both departments are responsible for specific production lines or cells; this minimizes the number of intersectional wafer and chip movements and thus the number of intersectional issues.

The use of continuous flow manufacturing also works to minimize the number of intersectional wafer and chip movements. For example, several wet-etch lines have been developed for specific chip types and the appropriate chemical baths are organized according to process sequence. Further, operators define the tasks of a process so that they can perform multiple tasks simultaneously. For example, operators who work on these wet-etch lines inspect each lot of completed wafers while the next lot of wafers is being etched.

Multifunctional Employees

Using multifunctional employees is an important aspect of product orientation. Operators are multifunctional in that they can perform most of the processes and functions in their section and sometimes in the entire department. About 60 percent of the workers in a section are qualified to perform any given process in that section. These operators inspect their own work, do simple maintenance, and work with the engineers to develop and improve their own procedures. For example, the furnace operators change the furnace tubes, run and check the experimental diffusion profiles, repair the boat loaders, and, as described shortly, improve their operations.

Other studies have found similar results. A study by the American Productivity and Quality Center found that workers in Japanese semiconductor factories have substantially more responsibility than workers in U.S. semiconductor factories. Japanese workers are responsible for the movement of lots, perform preventive maintenance, monitor equipment functions, evaluate results, and solve equipment and process problems. The study found that U.S. workers are responsible only for verifying lot identity, pushing the buttons to make a machine run, and calling a supervisor if there are problems.[1] Similar stories are found in other comparisons of Japanese and U.S. factories.[2]

Since Mitsubishi's multifunctional workers perform a number of activities that are performed by support staff in the United States, the use of multifunctional workers enables the company to have a smaller ratio of indirect to direct workers. For example, the wafer department needs only 75 indirect workers to support 360 direct workers. According to research by Warren Smith, a typical U.S. semiconductor factory has a larger ratio of support to direct workers. One U.S. factory, which produced only half as many wafers as Mitsubishi's wafer department, had 30 employees just writing specifications. The special engineers in this U.S. factory wrote the specifications; it was the job of the young process engineers to try to make the operators follow these specifications. Needless to say, improvements are difficult to implement in this type of atmosphere.[3] Other researchers have found similar results. Jeffrey Miller and Thomas Vollman found that manufacturing overhead costs represent 35 percent of manufacturing costs in U.S. firms, whereas they make up 26 percent of manufacturing costs in Japanese firms.[4]

Employee Developed Procedures

Employee developed procedures are an important part of the improvement programs in Mitsubishi's wafer department. The operators develop most of the process specifications themselves. They develop simple automation and more efficient procedures to reduce cycle time, inventory, contamination, and downtime. Other observers have also found that Japanese semiconductor factories emphasize the use and improvement of employee developed procedures. For example, one study concluded that the use of these procedures enables Japanese firms to develop higher quality semiconductor processes.[5]

In the wafer department, many of the improvements in the employee developed procedures are implemented via the suggestion system or small-group activities. Since many of the improvements are related to multiple steps in the wafer fabrication process, shared responsibility (another subelement of the product-oriented organization) is important. Responsibility for many of these improvements is shared between the operators in a working group and between the various working groups. And for improvements that involve their suppliers, the wafer and assembly departments share with these suppliers the responsibility for making the improvements.

Visible Management

Many of the improvements are related to visible management, the second major element of Japan's corporate culture. Problems are made visible by status boards and minimal use of solid walls. The wafer department is kept as open as possible to facilitate the identification of problems. Clear, Plexiglas windows are used so that operators can easily notice machines that have stopped operating, are leaking gas, or have other problems that may require intervention.

For example, there are no walls between the furnaces in the wafer department so that 4 workers can together operate 32 furnaces in 3 furnace bays. The operators have installed large display lamps on the furnaces that signal when there is a problem or when a process has been completed and operator intervention is necessary. Various meters and gas flow tubes have been installed on the front of the furnaces. These enable the operators to quickly check important process characteristics without going behind the furnaces.

Status boards are located everywhere in the wafer and assembly departments' factories. These boards display the cycle time for the wafer and assembly processes, equipment uptime, dust counts, pictures of the latest defects, and various types of quality data. For example, the photolithography section posts data on the uptime for each type of equipment and the number of reworks per week for various parts of the photolithographic process: spin, align, and develop. Major chip orders are also displayed on these status boards along with their progress through the wafer and assembly processes. As in every Japanese office, employee status boards also specify whether an employee is on-site that particular day and his approximate location at a given time.

Oral Communication

Oral communication, the third element of Japan's corporate culture, is also important to the performance of these semiconductor factories. As mentioned earlier, the factory layout typically matches the flow of the wafers and chips through the factory, which facilitates communication between the previous and subsequent processes in a factory. Workers often verbally ask the previous process for the wafer or chip types that they need to meet the requests of the subsequent process.

Further, the office layout for the support staff also matches the organization. This facilitates communication between the appropriate engineers and other support staff. Since the engineering section managers sit next to their section's engineering group leaders and engineers, these employees are constantly discussing recent problems or improvement activities as they receive the latest information from their factory or other Mitsubishi semiconductor factories. They hold daily standup meetings in which recent problems, other meetings, or improvement activities are summarized. This information is also quickly transferred to all three shifts via the operator group leaders, who also report to the engineering section managers. These group leaders meet daily with their section managers and provide their employees and the second shift group leader with daily updates. In a similar manner, the second- and third-shift group leaders pass on information to the subsequent shifts. In addition, the group leaders change shifts each week to improve communication between shifts.

A formal communication network also exists within the department and between the wafer and assembly departments and their major suppliers. The working groups mentioned above are one example of a formal communication network within the wafer department. They meet once or twice a month to discuss process improvements; their leaders meet once a month to discuss their groups' activities. In a similar manner, representatives of the wafer and assembly departments meet regularly, typically monthly, with representatives from their major equipment, parts, and raw material suppliers. They meet with representatives from the semiconductor equipment department to discuss existing equipment problems in the factory, the SED's new equipment, and the technology needed to make future generations of equipment.

The meetings with part suppliers typically focus on how to reduce Mitsubishi's costs or on the supplier's improvement activities. For example, with several other engineers from the semiconductor equipment department, I listened to a presentation by the assembly department's lead frame supplier. The supplier presented detailed data from over 50 separate experiments, its improvement activities, and its expansion plans. Employees of the SED said that they learned a great deal about the strengths and weaknesses of their suppliers and the suppliers' products through these presentations.

Extensive Training

The fourth element of Japan's corporate culture, extensive training, is used to introduce new employees to Mitsubishi's semiconductor factories and to upgrade their technical skills. In their first weeks on the job, new process engineers receive training on the company as a whole, the factory where they will be employed, semiconductor processes, technologies and equipment, and the factory's rules (e.g., handling of gas and chemicals) and procedures. This is followed by three weeks of on-the-job training, typically performed by one of the factory's group leaders. The group leader demonstrates the various processes that the new employees will perform. After a few months, they begin participating in small-group activities. The new operators are also assigned a mentor, who is expected to help the operators become a contributing part of Mitsubishi.[6] Operators continue to receive training throughout their careers. They typically learn new processes to better understand the entire manufacturing system. The best operators attend special schools where they learn about electronics and semiconductor physics. As described in Part Three, much of this training occurs when they are promoted.

CHAPTER FIFTEEN

Reducing Cycle Time:
Teamwork Among Engineers, Customers, Suppliers, and Operators

M ANY RESEARCHERS have found that Japanese factories emphasize reductions in cycle time. Mitsubishi's wafer and assembly departments are typical in this respect. They were able to reduce their combined cycle time from 72 days to 33 days between 1985 and 1989. This improvement is even more impressive in light of the dramatically increased product mix during this period. The number of chip styles that these departments produced increased to more than 700 in 1989, more than doubling between 1985 and 1989. The number of package types produced in the assembly department increased from 20 to 70 during the same period.[1] The reductions in cycle time were achieved through numerous engineer and operator activities. The engineers are primarily responsible for making system-wide improvements while the operators are responsible for implementing localized improvements.

The Roles of Engineers, Customers, and Suppliers

The engineers are concerned with the overall performance of the production system. With respect to cycle time, they are primarily concerned with the factory's layout, production bottlenecks, and product families. For example, engineers did most of the analysis for the development of several continuous flow lines including wet-etch lines, photolithographic lines, and the furnace cells. They also implemented continuous flow manufacturing by placing subsequent processes adjacent to each other. There are 17 different lines and cells in the wafer department. A typical wafer visits several of these different lines and cells, sometimes several times, during its complete process. Although many of these lines require different clean-room conditions, whenever possible the engineers have located the subsequent processes adjacent to each other.

Identifying Production Bottlenecks

Engineers in the wafer department also analyze the factory's production capacity to identify potential bottlenecks in the factory. The wafer department has a sophisticated computer program that contains the factory layout, the wafer process charts, and the forecasted product mix (number of chips/chip style). This computer program is used to identify which types of equipment to purchase to avoid bottlenecks in the production system.

The forecasted product mix is the most difficult information to obtain. This information depends on shared responsibilities and formal communication networks. Responsibility for developing the product forecasts is shared between the wafer and assembly departments and their customers, and a formal communication network exists to support these shared responsibilities. In regularly held meetings, these organizations discuss the production capability of the wafer and assembly departments and the future product needs of the customers.

The wafer and assembly departments also depend on shared responsibilities and formal communication networks to obtain new equipment quickly. Since chip shortages often affect most of the semiconductor firms at the same time, these firms tend to order similar equipment at the same time in order to increase their production capabilities. Therefore, equipment delivery times can be very long. However, since the wafer and assembly departments have very close relations with the semiconductor equipment department and meet regularly with the SED to discuss these issues, they receive priority on any equipment orders. This close relationship enables the wafer and assembly departments to quickly obtain equipment during expansions in the global chip market. By maintaining their production capacity, they are able to reduce or at least prevent increases in cycle time.

Developing Product Families

Another way in which engineers help reduce cycle time is by defining standard product families. These product families are related by process, not by the chip's function; this grouping helps reduce setup time, paper work, and eliminate mistakes. Operators merely need to check the chip's family as opposed to a detailed run sheet before completing the correct process. Often, these standard processes only require a specific button to be pushed. By simplifying the overall production process, the use of these product families helps reduce the cycle time.

The wafer department has defined four levels of standard chip families. First, there are eight chip families in which each type of chip in a family is processed by the same sequence of lines and cells. Second, there are 22 subfamilies in which each type of chip in a subfamily uses the same stations internal to these lines and cells. Third, there are 100 sub-subfamilies in which each type of chip in the sub-subfamily uses the same process parameters at the same stations. Fourth, some chips require different process parameters at various stations internal to the lines and cells.

These chip families are constantly evolving as Mitsubishi designs new types of chips. Although Mitsubishi does continue to sell old chip styles to its long-term customers, it often steers the customers toward new chips in order to simplify its product line and thus its processes. The close relationships between the wafer and assembly departments and their customers make it easier for these departments to influence their customers in this way. As discussed above, these departments regularly meet with their customers to discuss the future product needs of the customers and Mitsubishi's most recently developed chips.

The Role of Operators

Operators are also involved with reducing the cycle time in the wafer and assembly departments. Each operator is in a working group that meets once or twice a month. The boundaries between these groups overlap significantly. The groups define and implement their own projects, but the leaders of these groups meet regularly to discuss and coordinate the projects.

In 1989, the working groups in the wafer department were given very specific goals. They were expected to reduce the wafer cycle time from 17.8 days to 13.6 days and a typical wafer lot's processing time by 1.2 hours. Each member of the working group is expected to make 25 suggestions per year. The engineers evaluate these suggestions; typically 60 to 70 percent are implemented. The more senior engineers also attempt to guide the groups toward the most beneficial improvements; they work with the operators to write the procedures for an improvement. For example, consider the furnace cell mentioned previously in the discussion of visible management. Although the engineers developed the design for much of the original layout, the operators made several improvements to the cell, including an improved queuing system, an improved loading system, and an improved system for delivering lots to the subsequent lines and cells.

Improved Queuing System

The improved queuing system is a good example of how employee developed procedures facilitate teamwork. The furnace operators developed an error-free procedure for logging-in and logging-out individual lots of wafers. The simple procedures enable these four operators to share responsibility for the operation of the 32 furnaces. Instead of making each operator responsible only for a few individual furnaces, as is often the case in U.S. factories, the four operators developed and continually improve a simple procedure for processing wafers through the furnace cell.

In the furnace queuing system, lots are placed on shelves with laminar flow (a type of airflow that prevents dust particles from settling on the wafers). Each lot's number, process, and status can be seen from a distance of 50 feet. Since the 32 furnaces occupy an area larger than 50 feet by 50 feet, this visibility is important; it enables the four operators to share responsibility for the operation of all of the furnaces. Laminated run cards are stacked according to the sequence in which the wafer lots will be loaded into the furnaces. As the lots are loaded into the quartz boats that carry them through the furnaces, the lot stickers are removed from each box of wafers and attached to a card that describes the lot's sequence of operations. As the lots are removed from the furnaces, the stickers are removed from the cards and affixed to the boxes into which the wafers are placed. These simple procedures enable each of the four operators to understand the status of each lot that is to be processed in any of the 32 furnaces.

Improved Loading System

The improved loading system is a good example of employee developed procedures and the use of simple automation. After analyzing and improving the procedure for loading wafers, an operator designed a simple gripper that enables the wafers to be rolled from a plastic box into a quartz boat. This way individual wafers do not have to be individually handled. The operators modified the furnace tubes, flaring them at the ends so that the wheels on the quartz boats will not hit the tubes as they are pushed inside them. They also developed a method for specifying whether a wafer boat is in the pre-process, in-process, or post-process stage. Operators attach an appropriately labeled plastic tag to the end of the push-pull mechanism that specifies the correct stage of the furnace operation. These tags reduce the number of times in which wafers miss a furnace operation or are processed through two furnace operations.

Improved Movement of Lots

An operator working group also analyzed the cycle time between the furnace cell and the subsequent process to reduce this cycle time. First, they collected data on the existing cycle time. They found that 98 percent of the lots had a delivery time of less than two hours, with 55 percent below half an hour. Second, they asked other furnace operators for reasons why lots were held up in the furnace area. According to these operators, most of the delays were related to problems with computers or measuring equipment being down or unavailable (e.g., data dumps or other people using the equipment). Third, they classified these responses into categories and identified the relationships between the categories. Finally, they proposed solutions to each of these categories. They were evaluating various solutions when I returned to the United States.

Numerous other small improvements were also made by the furnace operators. Boat holders are wrapped in tape to reflect the furnace's infrared heat and prevent the holders from becoming too hot to handle. Flow tubes are located behind Plexiglas walls so that operators can continually monitor them. Plastic tags attached to each furnace specify its process and settings. Signs on valves specify their purpose, and labels on flow tubes specify correct levels.

Improved Yields and Lower Costs Through Cycle Time Reduction

Shorter cycle times facilitate improvements in chip yield and chip cost by providing faster feedback on the process improvements that are made to increase yield. Shorter cycle times also reduce the number of errors and the amount of wafer handling by operators. Wafer handling leads to wafer breakage and contamination, two of the leading causes of poor chip yields.

The continual improvements in cycle times and yield directly lead to low chip costs. Although the semiconductor operations at the Fukuoka Works use less advanced equipment than that used at Mitsubishi's other semiconductor factories, it is far more productive than a comparable factory in the United States. In one comparison between Mitsubishi's wafer department and a U.S. factory, the Mitsubishi factory produced four times the number of wafers per direct labor operator, had fewer support workers per direct worker, had 5 percent more of its wafers complete the wafer process, and had one-fourth the cycle time.[2]

The same study revealed that this factory did not use more advanced equipment than a typical U.S. factory. Instead, it found, Mitsubishi made better use of its equipment. Each area of the factory has implemented and continues to implement improvements similar to those described for the furnace cell.

These improvements lead to shorter cycle time, higher yields, less wafer breakage, and higher production of wafers per direct worker. The multifunctional workers enable Mitsubishi to have fewer support staff. Since the direct workers perform many of the activities typically performed by support staff in a U.S. factory, the direct workers can determine which activities are most important and how to improve the efficiency of these activities.

CHAPTER SIXTEEN

Automation and CIM: Teamwork Between Equipment Producers and Equipment Users

A S WITH MOST Japanese companies, Mitsubishi invests heavily in new manufacturing equipment for its semiconductor chip and other factories. As mentioned earlier, many of these equipment purchases are driven by various capacity analyses that identify the bottlenecks in the production system. Production capacity is continually increased as Mitsubishi expands its production of chips for its existing markets and develops new chips for new markets. New equipment is used to produce the newest chips while older equipment typically produces those semiconductor devices that require the lowest process technology.

Mitsubishi can use old equipment since it produces a broad line of semiconductor devices. It can transfer older equipment from products that require sophisticated process technology to products that can use less sophisticated process technology. According to Tomisha Yamada, a manager in the power device department, the oldest equipment is used to produce power devices. Since power devices require less sophisticated equipment than MOS (metal-oxide semiconductor) or bipolar devices, equipment is typically transferred from the production of MOS devices, to bipolar, and finally to power devices. Even within these three categories, equipment is typically transferred from high-technology products to low-technology products. Therefore, Mitsubishi can use equipment much longer than firms that do not produce such a broad line of semiconductor devices.[1]

In 1989, Mitsubishi's most advanced semiconductor factory produced 1M DRAMs. All of the equipment loading and unloading and the wafer handling is performed automatically. Workers enter the wafer processing areas only to perform equipment maintenance. In addition to Mitsubishi's production of high-volume chips such as 1M DRAMs, however, it is also developing a strong capability in the production of lower volume chips. The wafer and assembly departments, described above, are a successful example of low-volume products.

In addition, Mitsubishi began producing a large variety of chips at a completely automated factory in 1989. The level of automation in this factory is similar to the factory that produces 1M DRAMs, albeit some of the equipment is more advanced. The new factory can handle different sizes of wafers, frames, and chips. For example, an advanced packaging system automatically attaches multiple chips to a frame using a die bonder, attaches wires from a chip's pads to a frame using a wire bonder, and forms a plastic package around the frame using a molding machine. Programmable wafer feeders, frame feeders (adjustable rails), bonding heads, and vision systems enable the equipment to handle the multiple types of wafers, frames, and chips.

This equipment was developed by Mitsubishi's semiconductor equipment department. The development and installation of this equipment is described throughout the next two parts. The successful development and installation of this equipment depended strongly on the close relationship between the SED and Mitsubishi's semiconductor business.

Mitsubishi started the SED in 1986 to increase the competitiveness of Mitsubishi's semiconductor chip business. The semiconductor business guarantees the SED a market in return for input into the SED's design and technology decisions. The equipment is developed to meet the special needs of Mitsubishi's factories, although these factories agree on standard and optional features to reduce equipment costs. Since the SED designs and manufactures a broad line of semiconductor equipment, it is able to spread the cost of developing several new technologies among a variety of different equipment types.

Japan's recently developed strength in the semiconductor equipment industry will most likely increase the competitiveness of its semiconductor chip companies. Previously, Japanese firms merely used the same equipment (equipment produced by U.S. semiconductor equipment firms) better than U.S. firms did. In the future, Japanese firms will have access to equipment that is superior to the equipment U.S. firms will have access to. As the next two parts describe, Japanese semiconductor chip firms receive the most recently developed equipment before U.S. firms get it, due to the close relationships between the producers and users of the equipment. Semiconductor equipment is delivered to the semiconductor factories as soon as possible to begin working out its bugs. Therefore, U.S. semiconductor firms will be competing with Japanese firms that not only have demonstrated excellence in using equipment, but in the future will have access to equipment superior to that available to U.S. firms.

This scenario is not confined to the semiconductor industry, however. As Part Six discusses, Japanese firms have used these strategic alliances in almost every manufacturing industry. Automobile companies have formed long-term relationships with their parts and equipment suppliers. Computer, telecommunications, and consumer-electronic firms have developed long-term relationships with their chip and other component suppliers, many of which are subsidiaries or are internal to their firm or economic group (*keiretsu*).

PART FIVE

Teamwork in the Office: Designing Semiconductor Equipment

Japan's success at new product development is well documented. Japanese firms reportedly can design automobiles in two-thirds the elapsed time and one-half the engineering hours required by U.S. automobile firms.[1] Japanese firms tend to introduce new memory chips, heat pumps, special purpose mechanical transmissions, copiers, air conditioners, and consumer electronic products faster than do U.S. firms.[2] We also know that Japanese firms use strategies different from U.S. firms. These strategies include an emphasis on incremental improvement, standardization, closer relationships with both customers and suppliers,[3] and the use of overlapping problem-solving and "heavy-weight" project managers.[4] Moreover, Japanese firms are more successful at implementing "concurrent engineering" than are U.S. firms.[5]

While we know that Japanese firms use different strategies, however, we know very little about how they put these strategies to use or about how the particular uses of these strategies affect development cost, development time, or product quality. This is probably due to the nature of the product development process itself. Unlike manufacturing, where equipment layouts, amounts of inventory, statistical process control charts, and rework stations tell us a great deal about manufacturing techniques and performance, the product development process is difficult to see. Rather, it exists inside strategic plans, procedures, and other documents; it grows and changes in meetings, informal conversations, and other human interactions.

The subtleties of new product development suggest that we need more detailed studies of how Japanese firms develop new products and new technology for these products. Although studies that compare multiple product development processes are needed to validate hypotheses, without detailed studies of how Japanese firms develop new products, multifirm studies may not address the most important hypotheses. This is particularly true in light of the extreme differences in culture and language that exist between the United States and Japan. Important issues may not be addressed simply because they do not fit within our existing paradigms of how new products should be developed.

This part takes an in-depth look at how Mitsubishi's semiconductor equipment department designs semiconductor equipment. It shows how Japan's corporate culture helps this department reduce the time and cost of development and implement many of the strategies mentioned above. Since most Japanese firms use these strategies and have a corporate culture somewhat similar to that described in Part Three, Japan's corporate culture can be identified as a major reason for the success of Japanese firms in new product development.

In making this type of generalization, it is important to recognize that some characteristics of semiconductor equipment are not common to all industries and products. Using a two-by-two model developed by Clark and Fujimoto (internal and external complexity), semiconductor equipment has a somewhat simple customer interface (i.e., low external complexity).[6] In contrast to automobiles and other consumer products, it is relatively easy to quantify the needs of semiconductor equipment; because there are relatively few customers, much of this information can be gathered directly from them. Chapter 17 describes how the department's design engineers work with the semiconductor factory's process engineers to define the specifications for new equipment.

Since semiconductor equipment is internally very complicated, however, it has characteristics in common with high-technology industrial products and high-volume consumer products such as automobiles and personal computers, and to a lesser extent consumer electronics, appliances, and photographic equipment. With respect to semiconductor equipment, hundreds of assemblies must be designed and thousands of lines of software code must be written. Chapter 18 describes how the semiconductor equipment department quickly translates customer needs into designed, manufactured, and tested equipment. Development time and development cost are key issues to the SED; Chapter 21 describes how the department continually reduces cycle time and development cost by improving the department's design methodology.

Semiconductor equipment is also a low-volume product that is produced with fairly flexible equipment. Unlike high-volume products such as automobiles and appliances, rarely is a new semiconductor equipment factory or a completely new manufacturing process for producing semiconductor equipment designed and developed as new semiconductor equipment is developed. Other types of manufacturing equipment and many high-technology products such as aircraft, aircraft assemblies, engineering workstations, instruments, process control systems, medical equipment, and dental equipment are similar in this respect. Manufacturing costs are still important to the success of these products; however, many of the activities associated with developing low-cost, low-volume products are somewhat different than those associated with designing low-cost, high-volume products. Chapter 19 describes how each equipment design group works with the equipment manufacturing section and suppliers to design inexpensive equipment. Standardization is also an important way to reduce the costs of low-volume products. Chapter 20 describes how different equipment groups in the department work together to standardize software, hardware, and mechanical technologies.

In summary, the issues addressed in Part Five are most applicable to the development of high-technology products. Although the issues are more relevant to industrial products that are produced in relatively low volumes, they also apply to complex products produced in comparatively higher volumes, such as computers, telecommunications equipment, and auto-mobiles. Further, the development of high-technology products typically depends on the development of new manufacturing equipment. In this respect, the issues addressed in this part apply to all high-technology products.

CHAPTER SEVENTEEN

Engineers as Market Researchers and Salesmen: Teamwork Between Firms and Customers

THE SEMICONDUCTOR equipment department works closely with its customers, in this case Mitsubishi's semiconductor business, to design new semiconductor equipment. Mitsubishi created the SED in 1986 to support its semiconductor business. This type of situation is not unique to Mitsubishi. Japanese firms typically develop long-term, highly integrated relationships between customers and suppliers of a product, particularly in high-technology industries in which product and process technology change rapidly. These relationships are typically internal to a firm, with another firm in the same economic group (e.g., Mitsubishi Group), or with a small supplier through partial ownership. Most Japanese electronics or automotive firms are highly vertically integrated and have close relations between each level of customers and suppliers.[1] Because there is little information available as to how these relationships actually work, some examples will be instructive.

In the case of Mitsubishi's semiconductor equipment department, engineers are primarily responsible for communication between the department and the semiconductor chip business. There are no salesmen, marketing personnel, or other special liaisons in the department. A small business group of two engineers and three clerical workers handles the accounting transactions between the department and its customers. The multifunctional engineers perform most of the activities ordinarily performed by sales and marketing functions. They interface directly with semiconductor process engineers to define the equipment needs and the best way to meet these needs.

In other words, responsibility for designing and installing semiconductor equipment is shared between the design engineers and the semiconductor process engineers. A formal communication network exists to support these shared responsibilities. As described in Part Three, this network includes five types of communication meetings, which are typically full-day or multiple-day meetings.[2] Semiconductor equipment design engineers from the SED, semiconductor process engineers (equipment users) from the headquarters of Mitsubishi's semiconductor business, and semiconductor process engineers (equipment users) from each factory attend these meetings.[3] The engineers communicate the outcomes of these meetings to their coworkers via their daily standup meetings and weekly group meetings and through informal conversations in the open office.

The primary purpose of the individual equipment, assembly equipment, assembly technology, wafer process equipment, and wafer process technology communication meetings is to define the needs of Mitsubishi's semiconductor business and the best way to meet those needs. Wafer and chip yields, equipment downtime, new chip or package designs, new technologies, and equipment produced by other companies are some of the topics discussed in these meetings.

There is a constant give-and-take between what is needed by the semiconductor factories and what can be done by the SED.

Equipment produced by other companies helps these engineers determine what can be done. Each time another equipment supplier produces a new equipment model, Mitsubishi's semiconductor factories purchase the equipment, evaluate it, and work with the SED to improve Mitsubishi's equipment.

There is also a constant give-and-take between the semiconductor factories. Although each factory has slightly different needs, Mitsubishi's semiconductor business as a whole benefits by having these factories identify needs common to all of the factories. This teamwork enables Mitsubishi to make those decisions that are best from a long-term systems point of view, as opposed to a short-term segmented view.

Teamwork and Standardization

One example of this teamwork is in the area of standardization. To reduce individual equipment costs, the SED and Mitsubishi's semiconductor business attempt to standardize customer needs and the methods of meeting those needs. Further, these meetings attempt to standardize material handling and computer control methods to facilitate the implementation of CIM in Mitsubishi's semiconductor factories.

Example: Modular Design

Standard equipment modules are one example of this strategy of standardization. Modules are defined that can be repeated within and between equipment types. The semiconductor equipment department and Mitsubishi's semiconductor business define these modules in the formal communication meetings. For example, the etching and chemical vapor deposition equipment use the same structure of standard modules, in the wafer process equipment communication meetings. These modules include a wafer input, wafer output, load lock, and two processing chambers. The wafer input and output and the load lock (a vacuum is implemented at this stage) modules are common to both the existing etching and chemical vapor deposition equipment. Further, as discussed in Part Six, it was also decided in these meetings to use the same modules in the future equipment. The two processing chambers can be used as etch chambers, deposition chambers, or after-etch treatment chambers. With the etching equipment, the material that will remain on the chips (e.g., silicon, aluminum) as a pattern is etched in the first chamber and the masking material (i.e., photoresist) is typically removed in the second chamber.[4]

This modular design reduces manufacturing and development costs and development time. The standard portions of the equipment can be produced in higher volumes, with the necessary customization added near the end of the production cycle. Further, the technology associated with at least three of the chambers can also be shared between the two equipment types and multiple customers, thus reducing development costs and time.[5]

Example: Defining a Standard Series of Equipment

A more detailed example of standardization and modular design can be found in the department's standard series of digital wire bonder equipment. This line of equipment is replacing the cam-driven wire bonder equipment that until 1990 was primarily used in Mitsubishi's semiconductor factories. It uses digital control as opposed to cams to drive the wire bonding head and the frame feeders. The conceptual design for this line of equipment was developed in the wire bonder equipment and the assembly equipment communication meetings. Some of the conceptual design work was also done in the assembly technology communication meetings, since one of the project's goals was to connect the wire bonder equipment with the previous assembly process (die bonder equipment) and the subsequent assembly process (mold machine) into an advanced packaging system.

These meetings defined one type of wire bonder that can be used as a basic building block for a variety of systems. In addition to the die bonder and mold machine, the basic wire bonder can be connected to a quality control station, a standard frame loading or unloading station, or other wire bonders. Depending on the number of wires to be bonded to each chip, multiple wire bonders are typically connected with one die bonder and one mold machine into an advanced packaging system.

In addition, the conceptual design for a standard "digitalization" package was also developed in the wire bonder equipment and assembly equipment communication meetings. This package upgrades an existing cam-style wire bonder into a digital wire bonder at a cost lower than the cost of a new digital wire bonder. Therefore, the new digital wire bonder is primarily used to add capacity to a semiconductor production facility, while the digitalization package is used to upgrade the existing equipment. Since the upgrade package used only certain aspects of the newly developed digital wire bonder, the incremental cost of developing the package was very low.[6]

The new digital wire bonder and the digitalization packages are both building blocks for CIM. Standard equipment types, standard frame loading and unloading stations, and, as will be shown later, a standard control system facilitate the implementation of CIM in Mitsubishi's factories. The definition of these building blocks requires teamwork between Mitsubishi's factories and the SED. The shared responsibilities and the formal communication network between these different organizations in Mitsubishi facilitate the definition of these building blocks.

Example: Standard Equipment Features

A third example of standardization and the role teamwork plays can be found in the lead process equipment. This equipment cuts and forms the packaged chip leads using a series of three progressive dies. Since each package type requires a different kind of lead frame, however, different dies were needed for each package type. Mitsubishi's semiconductor factories produced five distinctly different packages in 1989, and the three dies were different for each package type. This procedure was expensive and required frequent changeovers.[7]

Projections of Mitsubishi's future package production were continually discussed in the lead process equipment communication meetings and the assembly equipment communication meetings. Using these projections, the lead process equipment group redefined the dies to reduce the number of changeovers required. The conceptual design for these new dies was developed in the lead process equipment communication meetings through teamwork between the various organizations represented at these meetings. A single first die now handles all of the different chip styles. The second die has been standardized so that only two different dies are required instead of five. Only the third die needs to be changed for each individual package style.[8]

Another aspect of Mitsubishi's lead process equipment was also improved in these meetings at about the same time as the new dies were developed. Each of Mitsubishi's semiconductor factories had been using different storage media to transfer frames to and from the lead process equipment. A variety of magazines were used to unload the lead frames from the mold machine and load them into the lead process equipment. Pallets, tubes, and reels were used to transport the chips from the lead process equipment to the final test station. Therefore, the lead process equipment required special input and output devices (e.g., vibratory feeders) that could handle pallets, tubes, reels, and each type of magazine. The mold machine required special output devices and the test equipment required special input devices. Any transfer of chips between Mitsubishi's semiconductor factories also required time-consuming changeovers.[9]

Engineers from the semiconductor equipment department, the headquarters of Mitsubishi's semiconductor chip business, and each of Mitsubishi's semiconductor factories recognized the inefficiencies in this system. Due to the shared responsibilities and the formal communication network between the SED and the chip business, they were able to simplify and improve the system. In several lead process equipment and assembly equipment communication meetings, these engineers worked together to choose standard transfer methods in order to standardize the input and output of these equipment types. One magazine type was chosen for unloading the lead frames from the mold equipment and loading them into the lead process equipment. The finished chips are now loaded into one type of pallet; the pallets can now be stacked on top of each other and can be loaded easily into the test equipment. Equipment costs and changeover times for the mold, lead process, and test equipment have been significantly reduced as a result of the design changes. Further, these changes will facilitate the implementation of CIM through a simplified material handling system.

CHAPTER EIGHTEEN

A Standard Design Methodology: Teamwork Between Design Engineers

THE MITSUBISHI semiconductor equipment department translates its customer's needs into new semiconductor equipment using a highly detailed design methodology. Customer needs are continually defined and redefined in the individual equipment, assembly equipment, and wafer process equipment communication meetings. The SED uses its formal method to turn these equipment needs, particularly standard customer equipment needs, into equipment specifications, detailed designs for the equipment, and finally the manufactured equipment.

Several elements of Japan's corporate culture have a strong effect on the SED's design methodology. The most crucial aspect is employee developed procedures and plans. These procedures determine the order in which each design activity is performed, the documentation to be developed in each of the activities, and the type of teamwork in each activity. In essence, these procedures determine the cost and quality of equipment produced by the department and the time it takes to introduce a new product.

Some readers may find the following description of the SED's design procedures too detailed. It is included, however, to demonstrate Mitsubishi's emphasis on developing a standard, repeatable design process that each engineer understands, uses, and helps improve. This is the type of design process required to promote teamwork and to reduce the design cycle time.

This type of design process is not unique to the SED. According to an engineer who transferred to the SED in January 1989, it was in fact much less detailed than the design process used in his previous department.[1] Michael Cusumano has also documented detailed procedures used in Japan's largest computer firms to translate user needs into software code. He describes these firms as "software factories" because of their extensive use of procedures.[2]

The SED's design procedures are similar in many ways to those used by the firms investigated by Cusumano: they are lengthy, detailed, well understood, and — most important — an integral part of the design process. The department's procedures are described in a 10-page document that each engineer has read, received training on, and helped revise in various working groups. These procedures represent many years of continual improvements in which the SED has attempted to define a design process that can develop the highest quality and lowest cost products in the minimum cycle time. The achievement of these goals requires the development and continual improvement of a repeatable process that everyone understands and uses. Employee developed procedures are the heart of this design process.

Outline of the Design Methodology

Figure 18-1 is translated from this 10-page document. It outlines the department's equipment design methodology. The flow of design activities appears on the right side of the figure, with the documentation required at each step shown on the left. Rules have been developed for how this process is to be performed and how the documentation is to be organized. Figure 18-2 shows the outlines for many of these documents. Each step in the outline contains between one and five subpoints that are to be addressed in the order shown. Since the documentation is developed as the equipment is designed, this outline also defines the order of the design activities.

Starting at the top of Figure 18-1, the product plan is developed by each equipment group's senior designers through frequent consultations among themselves and with the equipment users in the equipment communication meetings. The product plan describes the product's needs, previous products, problem points, related projects, basic manufacturing process plans; it also forecasts potential problems. Previous equipment problems are analyzed, particularly drawings that were the source of problems. The problems and data describing these problems are analyzed using quality functional deployment. Quality functional deployment, first developed by Japanese firms, uses a variety of matrices to relate customer requirements to various design characteristics. The semiconductor equipment department uses these types of matrices to clarify the relationship between equipment problems and design decisions.

System Specifications

System specifications, the next step shown in Figure 18-1, describe an equipment's functional requirements and functional configuration at a macro-level using numerous block diagrams. They are developed in the same way as the product plan — by each equipment group's senior designers through frequent consultations among themselves and with the equipment users (i.e., semiconductor factories) in the individual equipment communication meetings.

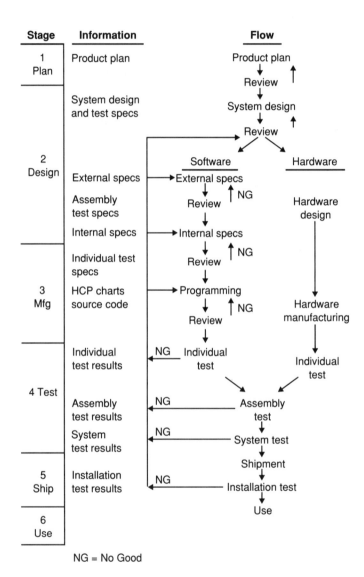

NG = No Good

Figure 18-1. The Semiconductor Equipment Department's Design Process

1. System Documentation
   a. Summary
   b. Scope of equipment application
   c. Structure of equipment
   d. Design conditions for each subassembly
   e. Equipment functions
   f. Output interface
   g. Functional configuration
   h. Program interface

2. External Specifications
   a. Summary
   b. Configuration of hardware and software
   c. Equipment block functions including I/O and error processing
   d. Program interface
   e. Operator interface
   f. Block interface
   g. Design conditions for each block

3. Internal Specifications
   a. Summary
   b. Processing method
   c. Common data
   d. Module configuration

4. Module Description
   a. Calling procedure
   b. Upper and lower modules
   c. I/O data
   d. Flags
   e. Function of module

5. Individual, Assembly, and System Test
   a. Summary
   b. Test requirements
   c. Test case
   d. Test procedure

Figure 18-2. Outline of Design Documentation

A number of documents are produced as part of the system specifications. The outline for these documents and their associated design activities are included in Figure 18-2. The order of these activities was chosen to help the design engineer follow a logical thought pattern from general equipment concepts to a detailed block diagram of the equipment. The *summary* section of this document describes the equipment's purpose and its functional outline. The *application scope* section describes those objects (e.g., frames, wafers, chips) that the equipment must handle and how the equipment handles them. The *equipment structure* section describes the subassemblies that will perform these activities. The requirements imposed on the equipment's subassemblies by the application are described in the *design conditions* section. These conditions are used to develop a block diagram (the *equipment functions* section) of the equipment's functional configuration, specifically focusing on the input/output (I/O) functions, I/O data, and error processing in this I/O function. The *output interface* section describes the I/O functions in more detail, including the equipment's timing charts, the signal structure of the related equipment, the data interface, and the human-machine interface.

The remaining sections of the system specifications describe the equipment's internal configuration in more detail. The *functional configuration* section separates the equipment into system functions and programs. For example, consider the advanced packaging system's quality control station mentioned as an example of standardization in Chapter 17. The quality control station is defined as one of the packaging system's "functions" and its control panel is defined as one of the station's "programs."

The functional configuration section includes a block diagram of each program's hardware and software. The purpose, function, I/O information, error processing, and performance conditions are also described for each individual program speci-fied in the block diagram. The relationships between each pro-gram are described in the *program interface* section. This section includes the operator interface and the software and data com-mon to multiple programs.

In summary, these design procedures force the design engi-neers to make an explicit connection between customer needs and the equipment's system specifications. Further, they force each engineer on the design project to use the same method of translating the customer needs into equipment's system specifi-cations. Since they are using the same method, everyone moves in the same direction and talks the same language.

Mechanical and Electrical Design Methods

After developing the equipment's system specifications, the equipment's design is still largely undefined. The equipment has been defined only at a macro-level in terms of its sub-assemblies and their functions. These subassemblies could be mechanical or electrical assemblies, although drawings or soft-ware code do not yet exist. The SED translates the equipment's system specifications into mechanical and electrical drawings, primarily through a team effort between each equipment group's mechanical and electrical engineers. This teamwork occurs at two levels. At the first level, the more experienced mechanical and electrical engineers work together to develop the equipment's block diagrams, timing charts, and logic dia-grams using the system specifications.

At the second level, the experienced engineers work with the younger engineers to develop the assembly and part drawings. The experienced mechanical engineers develop timing diagrams and rough sketches of the assemblies. The younger mechanical engineers, who sit at drafting tables, develop the detailed mechanical drawings based on these rough sketches and the information provided in the system specifications.[3] The more experienced electrical engineers develop rough sketches of the equipment's hardware and software including logic and block diagrams. The younger electrical engineers develop the detailed logic diagrams and timing charts, and detailed block diagrams of the hardware and its relationship with the software.[4] These block diagrams are used to develop the software's external specifications, which are described shortly.

These mechanical and electrical design activities are part of an iterative process involving constant consultation between the electrical and mechanical engineers and the younger and older engineers. Several elements of Japan's corporate culture facilitate this iterative design process. The open office, the daily standup meetings, shared responsibility, and extensive training have the largest effect.

The daily standup meetings and the open office facilitate communication between these engineers. Equipment problems and the status of individual projects are summarized in the daily standup meetings. Most of these engineers also sit next to each other in the open office. They are continuously asking each other questions and pointing out potential problems with the interface between the mechanical and electrical design. Although the young mechanical engineers sit at drafting tables, which are not with the rest of the equipment group, their desks are located in the same room and it is still easy for them to consult with the more experienced engineers. All the engineers can see each other from their desks and drafting tables, and if an engineer is not at his desk or drafting table, it is easy to find him by using the employee status boards.

Shared responsibility is an important part of these design activities. An explicit set of relationships between the engineers in an equipment group determines how responsibilities are shared between them. An example of these shared responsibilities was described in Part Three for the wire bonder equipment group. Typically, 12 engineers share responsibilities for a particular type of equipment. The sharing of responsibilities between a small number of engineers facilitates the early identification of design problems. Any one of the engineers might spot a potential problem, and in the open office, it is easy to notify the rest of the equipment group.

Extensive training is also an important part of this type of design process. The process provides the young mechanical engineers with a form of training and enables the older engineers to be involved with higher level engineering. Although this type of work breakdown also exists in U.S. firms, there are some important differences. In a Japanese firm, this work breakdown is part of an explicit long-term mentor and student relationship (described in Part Three). The younger engineer learns detailed mechanical design methods from his mentor, who sits near him, and he becomes involved with higher level design work as he gets older.

Software Design Methods

The software design process depends more on employee developed procedures and plans than does mechanical or electrical design. As can be seen from Figure 18-1, software design includes an additional formal step between system design and assembly test. The system specification is more explicitly connected to the computer code than to the mechanical and electrical drawings. Each step in this explicit connection requires the development of several new documents: In the case of mechanical and electrical design, on the other hand, no documents are required between the system specifications and the drawings.

There are two reasons for this explicit connection between the software code and the system specifications. First, since software languages follow a set of detailed artificial rules, it is much easier to define an explicit connection between higher level system parameters and software code. Second, since it is harder to visualize the operation of a software program, an explicit connection is needed more in software design than in mechanical or electrical hardware design to avoid operational problems.

External Specifications

As shown in Figure 18-1, the software design process involves the development of external specifications, internal specifications, and software code. The external specifications describe an individual program's block diagram (a set of boxes connected by lines), the relationship between the program and other programs, and the relationship between the various blocks in the program. A program is developed for each major block in an equipment block design. For example, the quality control station mentioned earlier represents an individual program in the advanced packaging system. Therefore, external specifications are used to document its software and the relationship between its software and its hardware.

An outline of the external specifications is shown in Figure 18-2. As with the system specifications, the documentation leads the designer through a logical sequence of activities that have been developed through years of continual improvement in the semiconductor equipment department's design procedures. The *configuration* section describes the structure of the program's hardware and software and the relationship between the hardware and software, via block diagrams.

The *functions* section describes the functions of the specific program, and the functions, I/O interfaces, and error processing methods used in each block in the program. Using information about each block's function, the software engineer also defines the relationship between the specific program, other programs, and the equipment operator. The *program interface* section describes how the specific program interrupts or uses other programs in the specific equipment. The *operator interface* section describes how the equipment operator would use the specific program.

The two final steps in developing the external specifications are to define each block's common software, data, and functions. Interfaces, I/O information, subroutines, tables, files and other resources common to multiple blocks in the program are specified. Each block's functions, I/O methods, error processing, and program and operator interface are also described. This information facilitates the use of common software, which is a powerful method for reducing software costs.

Internal Specifications

The internal specifications, the next step in the software design process, document a specific block within an equipment's program. For example, the quality control station's control panel represents one block within the station. Its software is developed and documented using the internal specifications. The *processing method* section outlines the algorithm to be solved by the specific block. The *common data* section describes the data common to multiple modules within the block. The *module configuration* section describes the relationships between the modules inside the block.

If a block within a software program is complicated, its modules are also described using a formally documented method. The term *software module,* however, has no formal definition in the SED. Basically, it can refer to any set of connected software lines that are self-contained (e.g., a subroutine) and are within one block. As described in Chapter 20, the semiconductor equipment department tries to use the same software modules in different software blocks and the same programs in single and multiple types of equipment in order to reduce software costs and increase software reliability.

The detailed software design procedures described here facilitate the development of standard software that can be used in multiple applications. The external and internal specifications make it easy for engineers to understand the potential for using programs developed for one equipment type in another equipment type. The documentation developed for software modules is even more powerful since it is easier to apply software at this level in the equipment's software structure to multiple equipment types. Therefore, the documentation for software modules was designed to facilitate the evaluation of a module's applicability to different equipment types. This documentation describes the method of using the module (i.e., calling procedure), the other modules connected to the specific module, the data needed to use the module, the data produced by the module, the type of flags used by the module, and the software flow internal to the module.

Hierarchical Control Process Charts

The internal specifications are also used to develop hierarchical control process (HCP) charts. These charts, which describe the information flow within the software, are written on large sheets of paper, using a formal procedure. Eventually, the software code is written next to the flow charts so that it is easy for engineers to understand the relationship between the actual software code and the software's flowchart. The code internal to repeated loops in the software is indented to clarify the beginning and end of these loops.

Nippon Telephone and Telegraph (NTT) developed this system, which is completely independent of the software language being used. The system is widely used within Mitsubishi and other Japanese companies.[5] HCP charts are somewhat similar to the flowcharts used in the United States, although they are much more detailed. Since they are developed before the software code is written, other software engineers can use them to integrate their programs with the programs described in the HCP chart before the actual code has been written.

Although the system developed by NTT contains about 20 different symbols, the SED uses only 6:

1. regular processes such as mathematical formulas, data transfers to and from memory locations, or definitions of variables
2. double branches such as a "yes" or "no" in an IF Statement
3. multiple branches such as the switch statement in the C language (the software language used in the department)
4. calling a subroutine
5. the end or beginning of a process
6. A repeated process such as a DO loop in Fortran or a FOR statement in C

For example, one symbol represents a repeated process. A description of the variable that is incremented in a DO loop or a FOR statement is written next to the symbol. The symbols (i=1; i=10; i++), for example, signify that the loop is performed from i=1 to i=10 in increments of one. The actual processes that are repeated are written below the symbol and these repeated processes and any processes internal to a branch or a loop are indented for clarity.

Advantages of the Software Design Process

This entire software process is quite labor intensive since it is all written by hand. Engineers spend a great deal of time writing and rewriting the specifications, particularly the HCP charts. Some observers might say that this time-consuming method portrays Japan's inability to write software. However, when the process is viewed from a systems point of view, the advantages become clear.

As described in Part Two, the first step in reducing design cycle time, reusing software code, and improving the quality and efficiency of the design process is to define and characterize the design process. The design process just described and its associated documentation are major steps in that direction. The SED's design process can be used by any member of the department. According to several members of the department, engineers can perform a step in the design process even if they don't perform the previous step in the process. For example, any of the software engineers within the wire bonder equipment group can write software code using another engineer's HCP charts (see Figure 10-4). In the same manner, any software engineer in the wire bonder equipment group can also develop internal specifications from someone else's external specifications or develop an HCP chart based on another engineer's internal specifications.[6]

This detailed design procedure facilitates the integration of different software blocks, programs, and equipment types by making it easy for the software engineers to understand the entire system as opposed to only specific sections of software. Since the software is stored in public files, it is accessible to everyone. As described in Part Two, software integration is very important because the number of lines of code in a typical software project has increased dramatically and is expected to continue growing in the future.

Further, this methodology also provides the SED with great flexibility in assigning individuals to various projects. This is important when trying to meet multiple time schedules. Frequently, a different engineer performs each step in the design process, depending on the work load for the various software engineers. Clerical personnel often do many of the final steps in the software design process, such as writing code from HCP charts and inputting the code into a computer. Although these people do not have college degrees, they typically receive short training courses on the design process and the C language.

Other elements of Japan's corporate culture also facilitate the SED's software design process. As described in Part Three, the software engineers typically prepare monthly schedules using standard forms, an aspect of visible management. Important software projects are described in detailed work schedules that assign an individual to each of the steps in the above design process on a daily level (see Figure 10-5). The open office also helps the engineers constantly consult with each other during these projects. Not only do these engineers sit adjacent to each other in the office, the computer room is organized so that they can also sit next to each other when they are working on-line.

Training is obviously important to the success of this design method. All new engineers receive about 15 hours of training on this design method, particularly the software documentation. Each new engineer is expected to read the various documents that describe these design methods; a new engineer's mentor is also an important source of information.

Procedures for Developing Test Plans

The semiconductor equipment department also develops equipment test plans as the equipment is designed, rather than after the design is completed. As mentioned in Part Two, it is important to perform these activities concurrently to ensure that a product's important performance parameters can be tested easily. Further, there should be a logical hierarchy of tests that will identify potential problems very early in the design process.

The SED has developed a set of design procedures that require the design engineers to concurrently develop the test procedures and the equipment design and to develop a hierarchical set of tests. As shown in Figure 18-1, the actual design steps, particularly the software definition phases, are inherently linked with the manufacturing and test steps. Each step in the design process requires the appropriate test specifications to be concurrently developed. There is a hierarchy of equipment tests from full equipment operation to individual assemblies and individual software modules. These test plans, in addition to software development plans and expected equipment operation methods, are documented as the system specifications, external software specifications, and internal software specifications are developed.

As Figure 18-1 indicates, individual tests are performed on a specific software module or block, electrical assembly, or mechanical assembly. Individual software test plans are begun while the software's internal specifications are being developed (see Figure 18-2). The *test requirements* section describes the method and type of tests to be performed in a specific software program, block, or module. The *test case* section describes the specific programs, blocks, and modules that are to be tested. These are identified using the appropriate internal specifications, module descriptions, and HCP charts.

The assembly tests are performed on individual electrical and mechanical assemblies. The plans for these are begun while the external specifications are being defined (see Figure 18-2). The *test requirements* section includes the equipment's performance objectives, the hardware and software structure needed to execute the test, and the method of testing the equipment. The *test case* section describes those programs and program blocks being operated during each specific test. The *test procedure* section describes the procedures and expected results for each specific test.

The system tests are begun after the equipment has been mostly assembled; they check the performance of the entire equipment's operation. The system test plans are begun while the software system specifications are being developed. These plans are very similar to the assembly test plans but include fewer details.

As the preceding discussion shows, Mitsubishi's semiconductor equipment department has developed detailed procedures (an aspect of employee developed procedures and plans) for designing its semiconductor equipment. These procedures represent many years of improvements in which successive problems with the design process have been identified and solved. Most of the design activities are performed in parallel to minimize the development time and to develop equipment that is optimized from a system cost and quality viewpoint. Test procedures are developed as the equipment is designed. The software design procedure reduces the number of software bugs by encouraging the software engineers to follow a logical set of steps from system specifications to software code.

Developing Inexpensive Equipment: Teamwork Among Design Engineers, Manufacturing Engineers, Suppliers, and Laboratory Personnel

THIS CHAPTER describes how the semiconductor equipment department develops equipment that is inexpensive to manufacture. When a new equipment model is developed, equipment manufacturing costs are considered at several stages in the design process shown in Figure 18-1. Basic manufacturing plans are developed in the first stage of the design process, the product plan. The manufacturing processes are then considered in more detail as the system specifications are translated into mechanical and electrical drawings.

Design engineers are expected to work with the manufacturing engineers and suppliers to perform cost analyses during this stage in the design process. The design engineers are primarily responsible for these analyses and in general are primarily responsible for an equipment's manufacturing costs. Since a large portion of manufacturing costs are determined very early in the design cycle, this strategy places the responsibility where it can have the greatest effect.

Sometimes called concurrent engineering, this strategy is not unique to Mitsubishi's semiconductor equipment department. Several observers of Japanese management claim that this strategy is more common in Japanese firms than in U.S. firms.[2] Chapter 19 takes an in-depth look at how one Japanese department implements concurrent engineering. As noted earlier, however, many of the activities associated with developing low-cost, low-volume products are somewhat different from the activities associated with designing low-cost, high-volume products.

Several elements of Japan's corporate culture play an essential part in the concurrent engineering in the SED. Its product-oriented organization locates design and manufacturing engineers together in the same department, with the department's office and factory located within 100 yards of each other. Although the design engineers are primarily responsible for equipment costs, this responsibility is also shared with the department's manufacturing engineers and its regular suppliers, since these organizations also participate in new equipment design. A formal communication network helps the department manage these shared responsibilities. One manufacturing engineer regularly attends each equipment group's weekly meeting. Further, as shown in Table 11-2, many of the wire bonder equipment group's design engineers spend a large fraction of their time in the SED's factory. Most of this time is spent performing joint experiments with the manufacturing engineers.

Manufacturing Cost Analysis of an Image Interface System

One example of how corporate culture helps the SED design inexpensive equipment is an image interface system used in the die bonder equipment. Mr. Hayashi, an engineer from the department's development group, was responsible for designing the improved image interface system. Working with two engineers from the die bonder group to define the specifications for the image interface system, he used a four-step process to design the system.

First, he analyzed the costs and functional capabilities of the three existing printed circuit boards and compared these capabilities to the desired cost and performance. Second, he used Mitsubishi's informal communication network. Due to the frequent interactions and employee transfers between the department and Mitsubishi's laboratories, the department could pinpoint the engineers at the manufacturing development laboratory who are Mitsubishi's experts in this technical area. These engineers suggested that Hayashi use a type of programmable gate array called a logic cell array.[2]

Third, since this array could replace a number of chips on the existing printed circuit boards, Hayashi aimed for a 50 percent reduction in costs and the combination of two printed circuit boards into one. Fourth, during the actual design of the printed circuit boards, he took advantage of the SED's close relationships with its suppliers. The new boards were designed and prototypes manufactured in five months. Compared to the previous system, the new image interface boards had lower manufacturing costs, a more compact design, and 30 percent faster processing speeds due to the more compact design.[3]

Manufacturing Cost Analysis of Mechanical Assemblies

Another example of how Japan's corporate culture helps the semiconductor equipment department design inexpensive equipment is in the design of mechanical assemblies. These are handled somewhat differently from electrical assemblies for a couple of reasons. First, mechanical assemblies are difficult to test before they have been manufactured. Mechanical assemblies are designed, tested, redesigned, and retested before they are ready to be produced regularly. Second, since new manufacturing systems and special tooling are not developed for new lines of equipment (due to low volumes), it is easy and inexpensive to modify the equipment designs. Therefore, mechanical assemblies may not receive a complete manufacturing cost analysis until many of these tests have been completed.

For example, the wire bonder equipment group introduced a new line of equipment in 1989, the advanced packaging system mentioned earlier. This equipment and many of its key mechanical assemblies had been under development for almost three years. Numerous tests were performed, initially by a laboratory, and subsequently by the manufacturing technicians under the supervision of the design engineers. Basically, the responsibility for performing these tests was shared between the technicians and the design engineers. The tests provided the design engineer with feedback concerning potential manufacturing problems long before the final design was developed.

However, not until after many of the key mechanical designs were verified during three months of production tests did the wire bonder equipment group allocate about 500 labor-hours to perform a manufacturing cost analysis and redesign of most of the wire bonder equipment's mechanical assemblies. These assemblies included the wire bonder's frame feeding system, the frame loading and unloading system, the enclosure for the wire bonder, the bonding head, and the wire bonder's controller.

The cost analyses are good examples of shared responsibilities in two ways. First, they were essentially a group problem-solving exercise involving several members of the wire bonder equipment group. The two youngest mechanical design engineers were responsible for the detailed cost analyses, with the senior designers contributing ideas. The young design engineers met weekly with the group leader, the assistant group leader, and the senior mechanical design engineer throughout a three-month period to discuss these cost analyses. Second, the suppliers participated in this process by contributing manufacturing process and cost information.[4]

The daily standup meetings were also an important part of these cost analyses. The wire bonder equipment group's leader summarized the cost reduction activities and asked the members of the group for cost reduction ideas almost daily for a couple of months. Numerous ideas were suggested as a result of these daily meetings.[5]

According to the assistant group leader, Kazuhiro Kawabata, group problem solving was an important part of this process. He claimed that each engineer has a different perspective on the wire bonder and individual engineers cannot be impartial. For example, the group leader has a global view, the assistant group leader has an electrical viewpoint, and the mechanical engineers, naturally, supply a mechanical viewpoint. Having several members of the group provide input to the design of these subassemblies increases the chances that problems will be avoided and that a good design will be developed.[6]

Three strategies are used by the wire bonder equipment group to reduce the equipment's costs.[7] First, they reduce the number of parts in each assembly. For example, the frame feeding system uses a set of rubber belts and steel rollers to transfer the frames underneath the bonding head, between different wire bonders, between the die bonder and wire bonder, and between the wire bonder and the mold machine. The wire bonder equipment group was able to reduce the number of parts by more than 50 percent in these assemblies through various group discussions and cost analyses performed by individual engineers.

Second, the wire bonder equipment group attempts to use lower capability components that can still meet the equipment's performance objectives. For example, the cost of a small motor depends on its positioning accuracy. Servo motors are the most expensive and accurate, followed by pulse motors and induction motors. The wire bonder now uses pulse motors for adjusting the width of the rails in its frame-feeding system.

When I left the SED, the wire bonder equipment group was planning to test induction motors to see whether they could handle the necessary positioning accuracy. Responsibility for this effort was to be shared between several of the wire bonder equipment group's design engineers and several members of the department's manufacturing section.[8]

The third strategy for reducing equipment costs is to locate a cheaper supplier of standard components. The SED depends on its informal communication network for a large percentage of these ideas. For example, standard optical components such as lenses, cameras, and illumination sources are used in the wire bonder's vision system. As discussed in more detail in Part Six, the wire bonder equipment group has worked and continues to work very closely with Mitsubishi's industrial systems laboratory to develop higher performance vision systems for position recognition of semiconductor chips. During a visit to this laboratory in 1989, the cost of these optical components was discussed in great detail while laboratory personnel helped both the assistant group leader of the wire bonder equipment group and me develop the conceptual design for a new vision system. Through these frequent conversations, the wire bonder equipment group remains familiar with the most recently available optical components, their costs, and their performance capabilities.

The Role of Suppliers in the Design of
Inexpensive Equipment

The role of suppliers in helping the semiconductor equipment department reduce its equipment costs has been mentioned several times in this chapter. Due to the close relationships between the department and its suppliers, however, the formal communication networks and shared responsibilities that support these relationships deserve more discussion. An exchange of employees is an important part of the formal network between the department and its suppliers. The suppliers often send engineers to the SED where they can design the subassemblies as the entire equipment is designed. For example, one engineer from Utsumi Electric, the department's major supplier of electrical assemblies, spent six months in the department helping the wire bonder equipment group design several printed circuit boards for the wire bonder's vision system. While designing these boards, he also studied the wire bonder's other printed circuit boards so that Utsumi Electric could improve its design and manufacture of them.

Another engineer from a supplier called Sekishin studied the die bonder for at least six months in 1989; he was still there when I left in December 1989. Although Sekishin was already making a variety of simple parts for the die bonder and other equipment types in the department, it wanted to start producing more complicated electrical assemblies for the department. Therefore, this engineer was studying the die bonder's control diagrams and PCB designs to identify potential business for Sekishin. He was involved with several die bonder development projects during his stay in the department.

A number of shorter visits are also part of the formal communication network. The SED's senior electrical and mechanical designers visit their suppliers' manufacturing facilities several times a year; the senior mechanical design engineer in the wire bonder equipment group made more than 10 visits in 1989.[9] The section manager responsible for the design of the wire bonder and die bonder equipment meets with the presidents of his major suppliers at least once a year to discuss general relations between his section and the suppliers.[10] These meetings typically last a couple of days and involve several members of his section. The other section members are briefed on these activities in the daily standup meetings.[11]

The heads of these small firms were even invited to the SED's end of the year party (*bonenkai*). These parties are important events for every Japanese firm. The fact that the department invited representatives of its suppliers to its bonenkai says a lot about the importance that the department places on these relationships. These parties are also a good example of how social events facilitate working relationships in Japan. The festivities included a beer and milk drinking contest; the winners of the beer-drinking contest (the objective was speed) then drank milk from baby bottles while blindfolded, where again speed was the objective. Several supplier representatives and I participated in the contest. Needless to say, after drinking from baby bottles in front of more than 200 guests, most of the participants, and even the observers, became fairly uninhibited. It was easy for the department's employees to meet representatives from each supplier in this situation.

Improving the Analysis of Manufacturing Costs

The department is constantly trying to improve its method of analyzing manufacturing costs. The desire for these improvements is driven primarily by the department's increased production of semiconductor equipment. As discussed earlier, the department is relatively young and has only recently begun developing a plan to sell some of its equipment outside of Mitsubishi. It realizes that its prices are now higher than its competitors, especially if it employs the sales and marketing staff necessary to sell equipment to non-Mitsubishi factories.

Therefore, the department is developing a more production-oriented "environment" for some of its equipment. Two of these changes involve its strategy toward manufacturing cost analysis. First, the department is trying to increase the amount of teamwork between the design engineers and the manufacturing section in the selection of a supplier for a specific part. For example, in 1989 the wire bonder equipment group's design engineers negotiated with both of its two major suppliers of custom parts and assemblies on each of the wire bonder's sub-assemblies and parts. However, the department's manufacturing section actually chose the specific supplier for a particular component or assembly *order* largely based on delivery time. Therefore, the procurement costs were sometimes higher than necessary. The department manager wanted to improve this process. He wanted the design engineer to work with the manufacturing section (i.e., shared responsibilities) to specify only one supplier for each subassembly as the equipment is designed.[12]

Second, some of the equipment groups are increasing the amount of responsibilities to be shared by the design engineers, the suppliers, and the manufacturing personnel in the design of new equipment. Plans were being developed to include manufacturing engineers from both the SED and the suppliers in the group-problem solving activities in the wire bonder equipment group. According to the assistant group leader, Kazuhiro Kawabata, most of the experienced design engineers in the department had previously used this type of method before they joined the department.[13] As part of these changes, in late 1989 a representative of the manufacturing section began attending the weekly wire bonder equipment group meeting to increase the manufacturing section's input in the group's design decisions.

CHAPTER TWENTY

Standardization: Teamwork Between Different Equipment Groups

THE SEMICONDUCTOR equipment department also develops standard software modules, electrical hardware, and mechanical technologies and applies these standard units to different equipment types to reduce equipment costs. As mentioned earlier, the definition and use of standard components, assemblies, and modules is an important strategy in most discrete parts manufacturing industries, particularly in the production of low-volume products. Because semiconductor equipment is a low-volume product with high development costs, standardization can reduce equipment costs significantly.[1]

In addition, the department believes that standardization helps improve equipment reliability and quality. The high yields required in semiconductor manufacturing make equipment reliability extremely important. Although the department has found that equipment reliability can be improved over an equipment model's life, short product life cycles are also characteristic of the semiconductor equipment industry. New models are introduced just as reliability begins to improve on the old ones.[2]

The department's number-one reliability problem is software quality and developing the means of measuring, controlling, and improving it. According to Hiroshi Honda, the wire bonder equipment group leader, this is also a problem with other Mitsubishi products (including high-volume products such as its VCR); software reliability was a major topic at an internal Mitsubishi technical conference in 1989. Since a new product's software typically builds off previous software, reliability problems can be built in from the first version of the software. This is particularly a problem in the case of low-volume products for which it is often prohibitively expensive to redesign the complete software structure. Therefore, if there are bugs in the original software, the bugs may be built into future product models; these bugs can become bigger rather than smaller over the product's evolution.[3]

The most important software reliability issue, however, is related to the implementation of CIM in Mitsubishi's semiconductor factories. As discussed in Part Four, CIM is an important part of Mitsubishi's efforts to increase yields and reduce costs in the semiconductor factories. The SED believes that standard hardware and software interfaces between different types of semiconductor equipment facilitate the implementation of CIM. It is working closely with Mitsubishi's semiconductor business to develop the type of standard interfaces that will facilitate the implementation of CIM in the company's semiconductor factories.[4]

This emphasis on standardization is not unique to the SED. It is a typical part of a Japanese firm's strategy. Abegglen and Stalk argue that standardization is a major reason for the success of Japan's manufacturing firms.[5] Other observers of Japanese management have made similar assertions.[6] Very little information has been published, however, as to how Japanese firms achieve this standardization.

Software Standardization

The SED's standardization efforts primarily emphasize the development and application of standard software modules, electrical hardware, and mechanical technologies. With respect to software standardization, about one-quarter of the SED engineers are involved with the development and debugging of software, primarily in those activities discussed in Chapter 18. Therefore, software costs represent an important percentage of equipment development costs.

Software Modules

Software modules that can be applied to multiple equipment types and multiple parts of one equipment type are seen as one way to reduce software costs and improve software reliability. The term *software module* has a variety of meanings, depending on the level in the software's hierarchy. In Chapter 18, the design methodology chapter, software structure is defined in terms of programs, blocks, and modules. However, even with this definition, software modules can vary in length; they can be defined in terms of HCP charts or programming, assembly, or machine language.[7]

When a new software module is developed, the department attempts to use or build from a previous module, whether the existing module is in the same equipment or a different type of equipment. This is particularly true when a new equipment model is developed. If the same software modules can be used, only new software interfaces must be developed. The identification of the appropriate software modules depends on the SED's design procedures, an aspect of employee developed procedures and plans. In particular, the appropriate software modules are identified via the external and internal specifications, HCP charts, module descriptions, and other software documentation described in Chapter 18. Frequently, the design documentation's summary of each software module provides enough information to determine if the software potentially applies to another design problem.[8]

For example, the SED standardized the frame loaders and unloaders used in each type of assembly equipment. The best software features of each loader and unloader were identified using the software documentation. These features were combined into one standard loader and unloader that is better than the previous ones.[9]

Existing software is also often used for another application via the open office and job rotation. The open office and job rotation enable most engineers to be knowledgeable about each other's work. Therefore, software engineers in the same group rarely duplicate each other's work. Further, engineers often use software that was originally developed for another type of equipment. For example, Koji Matsuda, a software engineer in the reactive ion etcher equipment group, benefited greatly from job rotation and the easy communication in the open office. He was temporarily rotated into the wire bonder equipment group. Through this assignment, Matsuda learned a great deal about the wire bonder's software. Through the open office, he kept in touch with the evolution of the wire bonder's software and was able to apply one of the wire bonder equipment's software modules to the reactive ion etcher. This module, which is over 1,000 lines in length, interrupts the equipment's controller and moves data from the equipment's memory chips to several logic chips on the controller. This software module could be applied to both types of equipment due to its general nature and because both types of equipment used the same control hardware.[10]

Software Libraries

Each equipment group in the SED also uses software libraries to facilitate the development of standard software modules.

Software libraries are a form of public files, albeit electronic ones. In addition to the standard mathematical functions available with most U.S. computing systems, these libraries also store those modules, in both programming and assembly language format, which appear applicable to multiple equipment types. For example, the wire bonder equipment group's library contains a listing of the wire bonder's main software modules and a more detailed listing of those modules that apply most closely to other equipment types. Since portions of the wire bonder vision system software are used in the die bonder equipment, the vision system software is listed in its software library. The software in these libraries can be accessed by any equipment group.[11]

The SED is also trying to develop a department-wide software library, another good example of employee developed procedures and plans. A software library would make it easier for engineers to know and understand the types of software modules developed by other equipment groups. The software working group, which is trying to improve the software design procedures in a number of ways, is trying to develop an easy method of classifying the existing software modules to help software engineers identify appropriate modules. This group is reviewing the existing modules and using this information to develop procedures for how to define modules. These procedures would be used to register software in the library as it is developed; to classify the module properly, they would ask several questions as the software is entered as a module. Eventually this system would provide a formal method of applying software modules. It would complement the "word-of-mouth" method, which also seems to work well in the department due to the open office and job rotation.[12]

As mentioned in one of the examples of using standard software modules, the same electrical hardware is a necessary prerequisite for applying the same software to multiple equipment types. In the fall of 1989, one of the software engineers said that, on average, 20 percent of the software used in each equipment type was common to most of the equipment.[13] This was primarily because many of the different equipment types used the same type of controller. Further, much of this standard software performed high-level functions, such as input/output, graphics, and console display, that are independent of equipment type. The engineer claimed, however, that once a new standard controller was developed, 50 percent of the software would be common to all equipment types. The SED expected this accomplishment to significantly reduce equipment costs and the costs of implementing CIM in Mitsubishi's semiconductor factories and to significantly improve software reliability.[14]

Standard Equipment Controller

Until 1989, the semiconductor equipment department had used more than one type of controller in its equipment, although all were based on the Mitsubishi series of programmable controllers. However, the department wanted to develop its own standard controller to reduce costs, improve equipment reliability, and facilitate the implementation of CIM. A key part of this strategy was to match the controller's design with the needs of its equipment and the needs of the factory control systems used in Mitsubishi's semiconductor factories. The department's objective was to perform a larger percentage of the controller's computations and the factory control system's computations using more standard hardware, more standard software, and more hardware as opposed to software, thus reducing the cost of implementing CIM.[15]

Consider the structure of the eventually developed standard controller and its relationship to a semiconductor factory's control system. The SED wanted to define a *common application logic* for the standard controller that would be embodied primarily in hardware. This hardware and its associated software refer to tables and other data bases in which the data depend on the type of semiconductor manufacturing equipment and the type of chips produced. The application logic asks for the equipment's and the chip's ID numbers; the hardware uses these numbers to check the appropriate table or data base. Since only the data bases are dependent on the type of equipment and chips, common hardware and software can be used in each type of equipment and each type of factory.[16]

To develop this common application logic and embody it in hardware, a great deal of teamwork was needed between the SED and Mitsubishi's semiconductor business. The headquarters of the semiconductor business is responsible for developing the higher levels of factory control and the SED is responsible for developing the equipment's standard controller. Responsibility for developing the hardware and software used in the application logic is shared between the semiconductor business headquarters and the SED. A formal communication network supports these shared responsibilities. These shared responsibilities and the formal network make it possible to apply the same factory control technology to multiple semiconductor factories; a large fraction of this development cost can then be amortized over multiple factories.

The Role of the Development Group

The design of the standard controller is a good example of teamwork within the semiconductor equipment department and the role of Japan's corporate culture in this teamwork. The department's development group was primarily responsible for developing the standard equipment controller. Initially, the two members of this group, Mr. Hirayama and Mr. Hayashi, had sole responsibility for the project. Due to several problems, however, a working group was created that contained members of each equipment group. Thus, responsibility was shared between the development group and the equipment groups via a formal communication network. The evolution of this group and its relationship with the department's other working groups is an interesting example of how shared responsibilities and formal communication networks evolve to meet the needs of a Japanese department.

The department's training program also played an important part in the success of this project. Both members of the development group joined the department in early 1989. One member came from Mitsubishi's computer group, where he had been involved with developing control systems for Japan's train transportation system. The other member had come from the headquarters of Mitsubishi's semiconductor business, where he had been working on factory control systems for the company's semiconductor factories. Both engineers brought technological expertise in computer systems that previously did not exist in the SED. Both engineers received an accelerated departmental training program (described in Part Three), in which they learned about each equipment type and the main equipment technologies. This training program introduced them to the key members of the SED and provided them with a good start toward understanding the equipment's control needs.

The development group's first task was to understand the existing controllers and identify the good and bad attributes of each. The group used the conceptual design documents (in particular the system, external, and internal specifications) described in Chapter 18 as its first source of information. Based on these documents, the two development group members asked specific questions of key members in each equipment group.[17] The introductory training program introduced them to the key members of the department; the open office facilitated this information-gathering exercise.

However, this approach ran into a problem within a couple of months. The development group felt that key parts of several documents were written differently and thus it was difficult to understand the attributes of each controller.[18] This is surprising because, as stated earlier, the department uses detailed procedures to standardize the documentation of its software, electrical, and mechanical designs.

The identification of this problem and its solution are a good indication of the importance placed on employee developed procedures and plans by the department. The department expects its procedures to meet the needs of the design process. When a weakness is found with the design procedures, the department improves the procedures. In this case, due to the importance of this project, the department manager was involved with implementing the improved procedures. He asked each equipment group to rewrite their existing controllers' block diagrams in a standard manner so that the development group could better understand the design of each equipment's controller.[19]

Formation of a Working Group

To facilitate the development of this improved documentation, a standard controller working group was formed. Its original purpose was to provide the development group with any needed information and to make suggestions about the design of the standard controller. However, more documentation issues continued to be raised by the development group and the standard controller working group. These issues led these groups to begin driving the other departmental working groups, which were involved with standardization and documentation issues. Figure 20-1 shows these other working groups and how they were affected by the standard controller working group. Since the CAD working group was also developing a data base of standard parts and part molds to be used on the CAD machines, it was involved with the department's standardization efforts. The software working group was developing better methods to develop and catalog standard software. The quality control working group was developing better rules and methods for ensuring design and manufacturing quality; its origin was a request by the Fukuoka Works' manufacturing control department to standardize the quality documentation used by all the departments at the works. The production management working group was developing a computerized production system for receiving and managing customer orders and making payments to the department's suppliers. Therefore, its activities were connected with the software and CAD working groups because it was attempting to integrate this production system with the department's design methodology.[20]

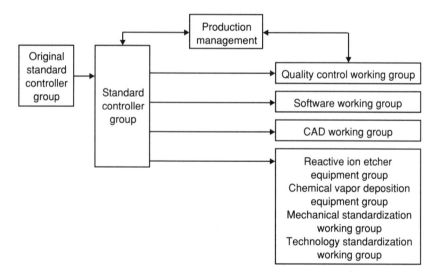

Figure 20-1. Relationship Between the Standard Controller and Other Working Groups

The standard controller group became involved with many of these other issues. In addition to developing technical standards for the standard controller, it had become the driving force in software and hardware standardization and design documentation. Most importantly, however, it recognized that the amount of standardization that could be achieved in the controller depended on the amount of standardization in the SED's mechanical technologies. Since a number of technologies and engineers had been brought to the SED from other Mitsubishi locations after the SED was formed in 1985, it still used a large number of different elemental technologies in its equipment. In many cases very different technologies were applied to similar problems or slightly different technologies were applied to the exact same problem. For example, various types of motors, air cylinders, and solenoids are used to drive similar mechanisms. Slightly different x-y tables and frame feeders were used in different equipment, although the same mechanism could be used in most cases.[21]

In general, the SED management believed that this technological diversity had produced excessive complexity in a type of product that is already inherently complex; for example, the wire bonder had about 35,000 lines of software code and far more than 10,000 individual parts. It had 2,000 individual electro-mechanical assemblies, consisting of more than 4,000 electrical components, 5,000 mechanical components, and myriad wires and connectors.[22] Therefore, the department decided to reduce the number of elemental technologies to achieve greater standardization of electrical hardware and software. Using standard mechanical technologies would basically reduce the number of mechanisms with which the standard controller would have to interface and would allow the department to concentrate on fewer technologies.[23]

Working Group Reorganization

The manner in which responsibilities were shared and the formal communication network that supported these shared responsibilities were changed in September 1989 to better meet the department's needs. The reorganization of these working groups reflects these changes in the department's shared responsibilities and formal communication networks.

Figure 20-2 outlines the new organization of the department's committees and working groups. The design committee included electrical hardware, software, and mechanical working groups; the actual committee members were the leaders of each working group. The hardware working group became responsible for electrical CAD and technology transfer of electrical technologies, and it reviewed the development of the standard controller. The software working group was responsible for the development of improved software documentation. The newly formed mechanical working group was created to develop standard methods of using CAD, define standard mechanical technologies, and set up a procedure for the transfer of these technologies between the three equipment sections. Each working group contained at least one member from each equipment group.[24]

At about the same time, the daily standup meetings were changed. Previously, each section held a standup meeting only on Monday, and each equipment group met separately on the other days of the week. After the working groups were reorganized, however, each section began to meet daily to improve the department's communication concerning the activities of the working groups. As described in Part Three, results from the working group's meetings were summarized in these daily section meetings. Following these meetings, each equipment group continued to hold its own standup meetings.

Office Work

General improvement activities	Design improvement activities
Engineering Section #1	Hardware Working Group • Standard controller • Electrical CAD
Engineering Section #2	Software Working Group • Software design procedures
Engineering Section #3	Mechanical Working Group • Mechanical CAD • Standard mechanical technology

Figure 20-2. New Organization of the Semiconductor Equipment Department's Working Groups

Standard Mechanical Technologies

The efforts to standardize mechanical technologies became one of the SED's key issues in late 1989. The department realized that the completion of this activity was a prerequisite for standardizing its software and hardware. Employee developed procedures and plans were an important aspect of these efforts. The software and hardware standardization efforts were held up until the mechanical working group could develop a procedure, as part of the department's design procedures, for applying a mechanical technology to a specific application. The mechanical working group met once or twice a week during November and December; its activities were summarized in the department's daily standup meetings.

However, the standardization of mechanical technologies as a strategy did not start with the design committee's mechanical working group. Just as developing a standard series of equipment is considered an important strategy in the SED (see

Chapter 17), standardization of mechanical and electrical technologies was always a key element of the department's overall strategy. A central part of this strategy is the identification of "key technologies": those technologies considered essential to the competitiveness of the department. These key technologies were described in one of the department's biannual strategic plans. This plan described how the department would standardize, maintain, continue to improve, and internally manufacture robotics, vision systems, x-y tables, loaders, unloaders, and feeder systems.[25]

Most of these standardization efforts were at the section level. For example, equipment engineering section #2, which is responsible for the design of the wire bonder and the die bonder equipment, held bimonthly meetings to standardize key technologies and components in the two equipment types. These meetings, which were attended by two or three senior members in each group, were attempting to standardize or continue using standard frame transfer methods, clamping mechanisms, position control (e.g., x-y tables), and other key electrical and mechanical parts (e.g., printed circuit boards). However, as the standard controller was designed, it become apparent that a departmental effort to standardize these mechanical and electrical technologies would have larger benefits than sectional efforts could produce. Therefore, the department reorganized the working groups and created the mechanical working group mentioned on the previous page.

The mechanical working group is an interesting example of how the product-oriented organization works, and in particular of how shared responsibilities evolve to meet an organization's needs. As described in Part Three, a major advantage of the product-oriented organization is that each equipment group in the SED performs most of the activities associated with that equipment type. It has responsibility for most of the decisions concerning its equipment type, and most of the time, each equipment group decides which technologies should be used in its equipment. In this case, however, the department decided to temporarily distribute responsibility for some of these decisions between the various equipment groups via a formal communication network — the mechanical working group. The mechanical working group had responsibility for choosing a standard set of appropriate technologies. Once it accomplishes this task, it will probably focus on other activities. Each equipment group will again become responsible for choosing its own technologies within the procedures developed by the mechanical working group. Over time, new technologies will become available and each equipment group may use different technologies for similar applications. At some point in the future, the department will probably form another working group to again spread the responsibility for choosing a set of standard technologies.

Returning to late 1989, the mechanical working group's main purpose was to develop a design procedure for choosing a technology for a specific application. First, it classified the department's mechanical technologies. The classification method, shown in Figure 20-3, characterizes these technologies by equipment type, function (type of movement), the object moved, and the type of actuator. This classification scheme enabled the working group to match the purpose of a technology (function and object moved) with the technology.[26]

Second, each equipment group was asked to classify each of their technologies according to this scheme, using a form developed by the working group. The form also requested that each equipment group supply a data sheet for each technology, including the cost, reliability, power, and other characteristics of the technology.[27]

This was the state of the department's standardization efforts when I left in December 1989. At that time, the mechanical

Type of Classification	Category
Type of Equipment	Wafer equipment
	Assembly equipment
	Molds
Function of Technology	Move up and down
	Move horizontally
	Rotate
	Move in 2 axes
	Move in 3 axes
Object Moved	Wafer
	Wafer cassette
	Magazine
	Frame
	Chip
	Expanded wafer
	Pallet
	Tablet
	Integrated circuit
	Tube
	Quartz boat
	Ring cassette
	Pallet cassette
Type of Actuator	Ball screw and motor
	Belt and motor
	Gear and motor
	Linear cam and motor
	Rotating motor
	Air cylinder
	Electric solenoid
	Liner motor

Figure 20-3. Mechanical Technology Classification Scheme

working group was planning to use the collected information to choose a standard method for each application and then adjust the existing technologies accordingly. Certain technologies would be authorized for specific applications. This information would be registered and made available to engineers in the form of a design procedure.[28] New employees will learn these design procedures when they enter the department, and as they learn about the department's major technologies, they will learn which technologies are appropriate for each application.

Improved Design Procedures to Reduce Cost and Cycle Time: Teamwork Between Different Equipment Groups

THE SEMICONDUCTOR equipment department continually improves its product design process. The existing methodology was described in Chapter 18; it includes the design of the equipment's mechanical and electrical assemblies and software. Documentation is developed at several stages in the design methodology, particularly for the equipment's software. This documentation includes system specifications, external specifications, internal specifications, and hierarchical control process charts.

The improvements in the department's design methodology and its associated procedures and documentation are driven by the department's needs. Some of these needs and their effect on improving the design procedures were discussed in Chapter 20. Chapter 20 described how some of the activities associated with standardizing software, hardware, and mechanical technologies identified weaknesses with the existing procedures and developed ways to improve these procedures. Most of these issues were related to the development of a new standard controller; the SED's development group had trouble, for example, understanding the configuration (i.e., block diagrams) of the equipment's existing controllers.

Throughout my stay in the department, however, design cycle time was the most important issue in the design improvement activities. Software is the last step in the design cycle and, as in most design processes, many of the design problems are solved in the final stages. Therefore, the software working group was the primary driving force in the improvement of the department's design process.

The Software Working Group

The software working group, which consisted of one software engineer from each equipment group, met once or twice a month in 1989 to discuss the software development process. The members of the working group each had at least five years' experience as software engineers and thus they had a very good understanding of the important issues. Relative to what they were able to accomplish, the members of this working group met very infrequently. The meetings were primarily used to reach consensus on what constituted the important issues, although the members of the software working group often could be seen in the open office discussing these issues.

The problems addressed by the working group changed several times during my stay in the department. These issues are described chronologically to provide the reader with a view of the dynamics of this improvement process and how the elements of Japan's corporate culture influence this improvement process. Until August, when the software working group released a report summarizing its past and future activities, it had primarily addressed:

1. the storage of software documentation
2. software procedures and documentation
3. software tools

Its future activities included software modules and quality management in addition to the above issues.[1]

Storage of Software Documentation

Software documentation is stored in the public files. These files contain the department's design documentation, experimental results, product and technology plans, and correspondence with customers and laboratories. Although this information was handwritten at that time, the department manager indicated that most of this information would be produced and stored on a computer by 1992.[2] However, to achieve this goal and to facilitate the use of the existing system, the department introduced an improved classification system for the public files in 1989. The department manager was very proud of this system; he claimed that this was one of his two best accomplishments in his first year as department manager.[3] His other major accomplishment, discussed in Part Six, was strengthening the department's equipment development capability.

The software working group was responsible for developing and improving the numbering system for software documentation and implementing a method of registering the software documentation. Although some of this work was completed as the improved numbering system was introduced into the SED, the software working group started a software subgroup (one of three software subgroups) in August 1989 to continue improving this system.[4]

The ultimate goal of this numbering system is to classify the software documentation in much the same way as assembly and part drawings are classified. And once this system has been developed, software can be ordered and delivered in a manner similar to the way assemblies and parts are ordered and delivered to meet a particular product option. Nonengineers will be able to prepare much of the software for a specific equipment order, since much of this work will be the combination of standard software modules.[5]

Improved Software Procedures

As described in Chapter 18, the SED uses very detailed procedures to develop software. Explicit documentation is required at several stages in the procedures. However, as described in Chapter 20, some sections of the software documentation were not standardized, making it difficult for other software engineers to quickly comprehend the documentation. During the summer of 1989, the software working group investigated each equipment's software procedures, considered adding or eliminating certain procedural rules, and redefined the range of issues to be covered in the high-level design information (i.e., system specifications).[6] For example, before the engineers can begin developing the configuration of the hardware and the software as required in the external specifications, they need information concerning the configuration of the input/output printed circuit boards and the memory map chips from the electrical engineers. This was a major issue in the development of the department's standard controller. Each engineer was documenting this type of information in a different manner. Therefore, the software working group developed a standard method of documentation.[7]

A major issue in this effort to develop better documentation is the ability to simultaneously develop the equipment's mechanical, electrical, and software design. Obviously, the mechanical design must precede the electrical design, and the electrical design must precede the development of software.

However, by understanding the exact information needed by the software and electrical engineer, both types of engineers can begin their design work earlier in the design process.[8]

Improved Software Development Tools

During the summer of 1989, the software working group also discussed the use of improved software development tools. It discussed the department's existing tools, commercially available tools, and tools being developed by Mitsubishi's information systems laboratory. By the time the software working group had produced a second report in November 1989, the software working group had developed a three-year, three-step plan to develop and implement a software development system capable of automatically producing software code and documentation from high-level inputs.[9]

First, the department planned to develop a software system that will enable the software documentation to be produced and stored on engineering workstations. This software is expected to organize the documentation according to the software numbering system discussed earlier; the department expected to have this system operational by mid-1991. As a first step toward this goal, the department purchased a package in 1989 that organized the existing software code and enabled code to be easily rearranged.[10]

Second, this software system was expected to produce software code from hierarchical control process charts by the end of 1991. As described in Chapter 18, an HCP chart is a standard type of flowchart that is independent of programming language. In 1989, Mitsubishi's information systems laboratory developed a software package that can translate HCP charts into software code. This package was developed as part of a corporate project called "Vision 88," the goal of which was to improve Mitsubishi's software productivity. Unlike typical U.S.

corporate-funded projects, this project was not a one-way technology transfer from the laboratory to various Mitsubishi factories. The SED and other departments within Mitsubishi evaluated the software developed by the laboratory and provided them with feedback as they were improving these tools. For example, Mr. Fukami of the SED worked closely with the laboratory in defining the corporation's software needs and developing these tools.[11]

Although the specific software package that produces software from HCP charts still had a few bugs, the department continued to test and improve this package in 1989. The package was operational, however, in the opposite mode: It could produce HCP charts from software code, which is very useful for producing software documentation. When changes are made in the actual software code, the HCP charts must also be changed to ensure correct documentation. The software package developed by the information systems laboratory can be used to automatically produce the correct HCP charts for the software documentation.[12]

Third, this software system was expected to automatically produce lower level software documentation, such as HCP charts from high-level documentation, by the end of 1992. This system will depend on the software numbering system described earlier and on the availability of software modules. Nonengineers will be able to use this system to prepare the software for a specific equipment order by combining appropriate software modules.[13]

Improved Software Quality

Software quality was also addressed by the software working group, while the numbering system, software procedures, and new software tools were being discussed in the summer of 1989. The chair of the software working group felt that quality was the most important issue concerning software productivity for at least three reasons. First, the quality of the high-level information (i.e., system specifications), which defines the interface between the mechanical, electrical, and software, has a strong effect on the equipment development's cycle time, because the quality of this high-level information determines how easy it is to translate the mechanical design into an electrical and software design.[14]

Second, software reliability is a major issue in the department; equipment downtime, which is primarily due to software bugs, is an important measure of equipment performance.[15] Third, most of the problems associated with the design process surface in the last stage of the design process, software production. Therefore, software problems tend to represent the sum of all design problems, including mechanical and electrical problems. These problems not only reduce equipment quality, particularly equipment reliability, but also delay the introduction of the new equipment.[16]

However, until the other issues, particularly software procedures, had been discussed, the software working group had difficulty defining the software's quality problems. After it had discussed these other issues and the standard controller working group had raised additional issues, it began classifying and describing software problems. It identified three types of problems that occur at different stages in the design cycle. The first and most serious type of problem involves high-level mistakes in the system specifications: the interface between hardware, software, and mechanical design. These problems surface during equipment testing and are very expensive to fix. They require the equipment designers to return to the systems design stage of Figure 18-1 and redesign much of the equipment.[17]

The software working group's solutions to this problem reflect the SED's emphasis on employee developed procedures and plans. The working group believed that better procedures were needed to develop the high-level documentation and to register and catalog software modules. For example, the standard controller working group identified a lack of timing charts in the high-level documentation as a problem with developing a standard controller. The software working group found that a lack of timing charts also causes high-level mistakes. Without these timing charts, the software engineers are solving a more complicated problem than actually exists. They attempt to develop software that satisfies *any* sequence of operations in a mechanical device, which often results in overly complicated software. Therefore, a standard type of timing chart is now required in the system specifications.[18]

The software working group also found that a better definition of equipment blocks, programs, and modules could reduce the number of high-level mistakes (the department's design procedures define equipment in terms of blocks, programs, and modules). Without the appropriate definition of an equipment's blocks, programs, and modules, it is difficult to test software, hardware, and mechanical designs until the equipment has been completely assembled. The importance of software modules was described in Chapter 20. Similarly, if blocks, programs, or modules that contain software, hardware, and mechanical elements can be defined and tested individually, problems can be located earlier in the design cycle. The department believed that its design procedures needed improvement in this area.[19]

The department was also planning to purchase a number of different simulators to help identify these high-level mistakes earlier in the design process. These simulators can test the equipment before it becomes operational. Software problems then could be identified very early in the design cycle, before any electrical hardware or mechanical assemblies had been produced. These simulators were expected to be used at a variety of stages in the design cycle. The level of detail used in the simulation models will depend on the stage in which they are used. The effective use of these simulators depends, however, on the existence of good software design procedures. Standard procedures are needed so that the appropriate simulation programs can be identified and developed. Without these procedures, it is difficult to identify the most important simulation programs to develop.[20]

The second type of software quality problem identified by the software working group involved the translation of external specifications into internal specifications, HCP charts, and software code. Under the existing test procedures, these mistakes often were not identified until software or equipment testing. Subsequently some of the higher level specifications had to be changed when these problems were identified. The software working group believed that the solutions to this problem were similar to the solutions just described for the high-level problems: improved procedures and standard software modules. More detailed design procedures promote greater standardization in the way the external and internal specifications are developed. This reduces the chance of misinterpretation as the external and internal specifications are translated into the software code. Standard modules will require less development of custom software and thus reduce the number of problems that need to be solved.[21]

The remaining problems identified by the software working group included coding and operation mistakes. Coding mistakes are made when the actual software code is produced from the HCP charts. Operation mistakes are keystroke errors. The software working group believed that coding mistakes also could be eliminated through more standard procedures and through automation. The standard procedure in this case included the identification of standard software code for each type of symbol in an HCP chart.[22]

Mechanical and Electrical Design

Employee developed procedures and plans and other sub-elements of Japan's corporate culture are also evident in the SED's improvement of its mechanical and electrical design activities, particularly in the implementation of computer-aided design (CAD). The department emphasizes the development of standard procedures for using CAD. Responsibility for developing the procedures is basically shared between the equipment groups via a working group that includes one member of each equipment group. The working group's progress is summarized in the department's daily standup meetings. The working group and these standup meetings constitute a formal communication network that enables the working group to receive feedback from the other members of the department. It also enables the working group to develop the consensus needed to implement the procedures for using CAD.

The department used very little CAD equipment in 1989. It had only 10 CAD stations, which were rarely fully utilized, as compared to 36 drafting tables, which were almost always utilized. In the wire bonder equipment group, only the bonding head and the frame-feeding system had been designed using CAD. The CAD system could not produce tapes, only drawings; the part machining information had to be manually loaded into the NC machines.

Probably few major U.S. manufacturing companies have as poor a CAD capability as this department. This is probably not an isolated case; the market for CAD equipment in the United States is much larger than the Japanese market. Further, the market for computers in the United States is also much larger than the Japanese market. As described in Chapter 20, most documents in the SED are handwritten.

Why Japanese Firms Use Less CAD than U.S. Firms

Why would firms in Japan, arguably the most technologically advanced country in the world, use substantially fewer computers than the rest of the industrialized world? I asked this question of many engineers in the SED during 1989 and received a variety of answers; they can be classified into four categories.[23]

First, Japan has not emphasized those industries that are the best applications for CAD. It has a small defense industry, which is where CAD was first applied in the United States. Japanese technology has been driven primarily by its high-volume manufacturing industries, where design costs are far less important than in low-volume industries. Only in the last 10 years have Japanese firms in low-volume industries started to export products such as machine tools, robots, and semiconductor equipment.

Second, the Japanese language is difficult to computerize. Until recently, personal computers did not have the necessary memory or screen quality to handle the Japanese language. The Japanese language has over 2,000 characters, including two phonetic alphabets and Chinese characters called *kanji*. Most kanji characters are extremely complex; some incorporate more than 20 individual lines or brushstrokes and require many times the number of pixels needed to represent a Roman character on a computer screen. Further, entering a Japanese word into a computer requires a larger number of keystrokes than a comparable English word.[24]

Third, most Japanese have never used a computer or even a typewriter. Japan has never used typewriters due to the number of characters in the language. None of the engineers in the SED who were over 30 years old had used a computer before entering Mitsubishi. Even in 1989, most universities had very few computers.[25]

Fourth, the writing of Chinese characters, or calligraphy, is still considered an art. Many Japanese still feel that good penmanship is required of an educated person. A set of kanji characters drawn with a fine brush is often considered artwork; these pictures are often prominently displayed in Japanese homes, offices, or art museums. Therefore, many Japanese have trouble even considering the use of a computer, since it represents a break with a strong cultural tradition.[26]

However, these four reasons still do not explain the lack of CAD stations in the semiconductor equipment department. Given Mitsubishi's financial resources, the department could easily buy a number of CAD stations and have their design engineers use and become proficient at them. The department certainly has the desire to use CAD. Both of the department's 1989 semiannual strategic plans emphasized the use of CAD. The number of drawings produced by each equipment group over time and the percent utilization of the CAD equipment were prominently displayed in the office. According to the department manager, all of the drafting tables would be replaced by CAD stations, all drawings would be copied into an electronic filing system, the documentation would be created via computer, and the CAD equipment would be connected to the SED's mold equipment by 1992.[27]

The SED's Approach to CAD Implementation

However, the semiconductor equipment department did not approach this problem as a typical U.S. firm would. Instead of replacing the drafting tables with the CAD equipment and allowing each engineer to display his creativity on the equipment, the department wanted to define the best procedure for using CAD before it implemented CAD equipment on a wide scale. The CAD committee was defining a standard procedure for using CAD, a standard for how a drawing should look, and a data base of standard parts and assemblies to facilitate their use. As with the software documentation, the department wanted engineers to be able to understand a drawing quickly. The department felt that defining a procedure for developing a standard drawing was one of the methods to achieve this goal. As these standard procedures are developed, the department plans to train the engineers in these methods as it has trained people in the other standard design methods.

Does this procedural emphasis and the four problems cited above constitute a competitive advantage or disadvantage for the semiconductor equipment department, for Mitsubishi, for Japan? The difficulty of computerizing the Japanese language and the cultural barriers to using computers are certainly disadvantages, the effect of which is seen in the absence of computers in Japanese offices. However, even without computers, Japanese firms have been able to design products that meet customer needs, introduce new products quickly, design these products using concurrent engineering, and benefit from standardization. Japan's corporate culture is the primary reason for the success of Japanese firms in these aspects of product design.

Japan's corporate culture, particularly its emphasis on employee developed procedures and plans, will probably also help Japanese firms automate their design process, since Japan's ability to continually improve its procedures before automating those procedures is a distinct advantage. It seems to have been a distinct advantage in the factory. Since Japan has succeeded in automating its factories at a faster rate than the United States has, despite its disadvantages with computers, it may surpass the United States in design automation through its continual improvement philosophy.

Continual Improvement Philosophy

As described in Part IV, Japanese factories continually improve their performance in small incremental steps. Automation is implemented only after the fundamental quality and material handling problems have been solved. In a similar manner, the SED is improving the procedures used in its equipment design process in small steps via various group activities. Automation is applied only after the design procedures have been simplified and the quality problems are well understood.

Quality is emphasized in the design process in a manner similar to the way it is handled in a Japanese factory — design problems are identified, classified, and analyzed. These problems are identified not just by the software working group, but also in other activities like the department's efforts to develop a standard controller. Test procedures are developed at key stages in the design process. These tests are performed by the design engineers as opposed to specialists, just as production workers are responsible for quality in a Japanese factory.

Cycle time, like JIT manufacturing, is also heavily emphasized in the design process — identifying and eliminating the bottlenecks. In the design process, the bottlenecks represent information: for example, what design information is needed to begin software development earlier in the design cycle. Standard procedures have been developed and are being improved in order to begin writing "zero defect" software earlier in the design cycle.

Simple automation is also applied to the design process in a manner similar to the application of simple automation in the factory — one step at a time. This automation builds off the department's design procedures, just as automation in Mitsubishi's semiconductor factories builds off material handling procedures.

Consider a five-step scenario that seems to be occurring in the SED. First, the department has defined and continually improved its design procedures. Second, it plans to use computers to write and store the design documentation according to the classification system that was developed in 1989. Third, it will use computers to generate software code from HCP charts and HCP charts from higher level documentation in some cases. Fourth, CAD stations are being implemented using standard procedures. Fifth, the department will connect the conceptual design system, the CAD system and the manufacturing system into a computer-integrated manufacturing system.

PART SIX

Teamwork Between Laboratories and Factories: The Development and Application of New Technology

Japanese firms are becoming the world leaders at commercializing new technologies. VCRs, compact disc players, digital audiotape recorders, facsimile machines, printers, copiers, laptop computers, and high-performance automobiles are a few of the products in which Japanese firms have become the technological leaders.

This success with new technology isn't due only to greater spending on R&D, however. Many successful Japanese firms spend less on R&D than their U.S. and European competitors. NEC, for example, spent less on R&D than its U.S. and European competitors during the 1980s but grew much faster than its competitors during that period. Although in 1980 it was one-third the size of GTE and served similar markets, by 1988 NEC was 50 percent larger than GTE. When Canon passed Xerox in unit market share for copiers in 1983, it was spending only a fraction of what Xerox was spending on R&D for reprographics at that time. Since then, Canon's sales have continued to grow much faster than Xerox's because Canon has expanded into new markets.[1] And although Japanese automobile firms are receiving more patents each year than U.S. firms and are applying many technological advancements to their automobiles before U.S. firms, they spend less on R&D per vehicle than U.S. firms.[2]

We also know that Japanese firms use different strategies than U.S. firms. Not only do they emphasize incremental improvements in product and process performance more than U.S. firms, they are also more tightly coupled, with closer relations between their laboratories and factories and with their suppliers and customers. They also have closer relationships between related businesses in order to develop core technologies that apply to multiple businesses.

While we are aware of these different strategies, however, we know very little about how Japanese firms put these strategies to use or about how the particular uses of these strategies shorten the time to commercialize new technology or increase the output from R&D spending. As with the product development process, it is difficult to see how technology is developed and applied. The subtleties of the commercialization of technology suggests that we need more detailed studies of how Japanese firms develop and apply new technology.

Part Six presents a detailed look at how one Japanese firm develops and applies new technology. It describes how the elements of corporate culture help Japanese firms incrementally improve product and process performance and be more tightly coupled than U.S. firms. A key focus is the relationships between Japan's corporate culture, the incremental improvement of products, and the tight linkages within and between Mitsubishi's laboratories, corporate headquarters, divisions, suppliers, and related businesses. The strategy of incremental improvement and Japan's corporate culture work together to tighten the linkages. The incremental improvement of products does so by requiring more frequent transfers of technology from laboratories to divisions than does a strategy that focuses on radical improvements. Just as Japanese firms' emphasis on incremental improvements in their factories has led to the easy flow of materials through these factories, their emphasis on incremental improvements in their products has resulted in an easy flow of new technology from laboratories to divisions. And just as incremental factory improvements required extensive teamwork by all employees in a factory, the incremental improvement of products has required extensive teamwork along the linkages mentioned above.

I am not arguing that all Japanese firms develop and apply technology in the same manner as Mitsubishi. However, since Japanese firms are more likely than U.S. firms to use the strategies mentioned above and to have the corporate culture described in Part Three, a look at how corporate culture helps one Japanese firm implement these strategies will suggest how the culture might affect other firms. In addition, since many of the organizational issues associated with developing and applying new technology are independent of the type of product or technology, the issues discussed in this part are relevant to many U.S. manufacturing firms.

Chapter 22 describes how Japan's corporate culture helps Mitsubishi's laboratories, corporate headquarters, and divisions work together to define Mitsubishi's annual R&D budget. Many U.S. firms face this issue. The remaining chapters deal primarily with Mitsubishi's semiconductor business. The organizational issues covered here are most relevant to businesses that need close relationships with their customers, suppliers, and laboratories. Chapter 23 describes how Japan's corporate culture helps Mitsubishi's semiconductor equipment department work with the Mitsubishi laboratories and semiconductor business to define the technological needs of semiconductor equipment. Chapter 24 discusses how Japan's corporate culture affects the way the SED works with Mitsubishi's laboratories to develop new equipment technology. How the elements of corporate culture help the SED work with Mitsubishi's semiconductor business to install a new line of semiconductor equipment is the subject of Chapter 25. This part concludes with a discussion of the tight supplier-customer relationships that are common in Japan and how these types of relationships enable Japanese firms to take a longer term view than typical U.S. firms.

Integrating Development Projects and Allocating Corporate Research Funds: Teamwork at the Corporate Level

M ITSUBISHI USES a highly developed procedure to integrate individual technology development projects and allocate corporate funds to the development of new technology. By emphasizing the integration of individual projects, this procedure enables Mitsubishi to develop technology for multiple applications and perhaps to achieve a far greater return on investment from their research expenditures than do firms that do not integrate their individual projects. Since Mitsubishi contains a number of related businesses, most of its development projects have applications to multiple businesses. This situation is not unique to Mitsubishi. Unlike many U.S. firms, which have either diversified into unrelated businesses or are small and have very weak linkages with other businesses, Japanese firms tend to be well integrated, both horizontally and vertically. For example, many of Japan's electronics firms are in the semiconductor, computer, consumer electronics, telecommunications, and electrical energy businesses. These firms can apply the results from many of their corporate-funded projects to multiple businesses.

Mitsubishi's Corporate R&D Selection Process

The corporate R&D selection process at Mitsubishi is a good example of employee developed procedures and plans. Mitsubishi uses a formal procedure to identify and integrate R&D projects that are necessary to the growth of individual departments, works, businesses, and the corporation as a whole. This procedure is outlined in Figure 22-1. Beginning from the bottom of the figure, factories and laboratories propose projects, which can be funded from a variety of factory, laboratory, business headquarters, or corporate headquarters budgets. These projects are developed using a bottom-up approach. Each section of a factory or laboratory develops a research budget that is modified and integrated with other plans at the department, works, and laboratory levels. (A number of other Japanese firms have been observed to use this type of approach.[1])

Importance of Employee Developed Plans

Although factories and laboratories officially propose these projects in September of each year, many of this process's most important activities occur before the project proposals are actually submitted. The proposals are typically based on product development requests, on each business's one-, three-, and five-year development plans, and on numerous formal and informal conversations.[2] Product development requests, project proposals, and business plans are good examples of employee developed procedures and plans. Product development requests are used by Mitsubishi internally to communicate product needs. Project proposals are submitted by a works to a laboratory; these are often based on the product development requests.

Mitsubishi's power device department, for example, receives product development requests throughout the year from other Mitsubishi businesses that it supplies with a large number of power devices. According to Mr. Yamada, a section manager in the power device department, the product development requests

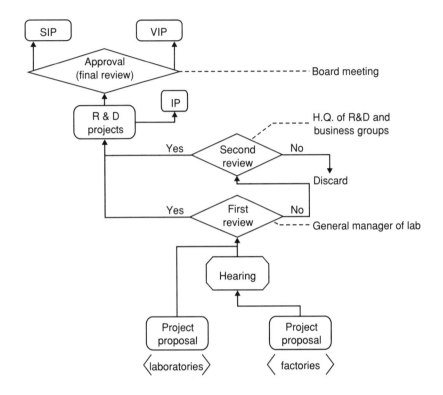

Figure 22-1. Corporate R&D Selection Process

frequently lead to sales of these power devices and help the department decide which technologies to develop and which projects to propose to the laboratories.[3] The requests are written on a standard form that includes the product's application (e.g., videotape recorder, automotive), the type of power device, the needed technological developments, the requested development date, and the proposed funding.

Mitsubishi's one-, three-, and five-year technology development plans also help Mitsubishi's businesses develop product development requests and project proposals. These plans are developed by each of Mitsubishi's major business segments, primarily through a bottom-up approach. Sections both in laboratories and in works develop technology development plans that are then integrated with the plans from other sections and departments. During the latter half of the R&D project selection process, these plans are written, distributed, and rewritten. They are distributed to a business's major factories, internal Mitsubishi customers, R&D laboratories, and corporate headquarters for comments before they are officially released. Based on these comments, the official plans are released in April (the beginning of Japan's fiscal year.)[4]

For example, each equipment group in the semiconductor equipment department writes a technology development plan for its equipment. These plans are then integrated with the plans from other equipment groups in the SED. Finally, they are integrated with the plans from other departments within Mitsubishi's semiconductor business. The 1989 semiconductor business plan was 175 pages; it included the market, basic plan, new developments, competitors, joint ventures, development schedule, world market, and sales for each of Mitsubishi's semi-conductor-related products.

These plans are very helpful in writing product development requests and project proposals. Departments within a single Mitsubishi business and other Mitsubishi businesses keep track of the technological direction and product needs of other departments and businesses through these plans. By under-standing the technological direction of these other departments and businesses, departments can submit product development requests to other Mitsubishi departments, and these depart-ments can subsequently submit project proposals to the labora-tories based on the product development requests.

Communication Networks and Integrated Product Proposals

Formal and informal communication networks also help Mitsubishi develop and integrate these business plans and use them to write product development requests and project proposals. With respect to formal communication networks, internal Mitsubishi conferences are held several times a year on a variety of technical subjects to transfer technology between different Mitsubishi businesses.[5] Short summaries of all R&D projects involving semiconductors and their related technologies are distributed to all Mitsubishi departments that are involved with semiconductors and their related technologies. Sections and departments that supply or purchase from each other or that produce related products also often have formal communication networks.[6]

Informal communication networks also help Mitsubishi develop and integrate these business plans and write product development requests and project proposals based on these plans. For example, Mitsuhito Watanabe, manager of the factory automation department in the manufacturing development laboratory, each month visits one or two works, receives visitors from the works between two and four times, and talks on the phone frequently with employees from the works.[7] Shinjiro Kawato, group leader for the vision group in the factory automation department, visits the works about once a month and receives visitors from the works many times each month.[8]

Other researchers have also found that Japanese firms have a significant amount of communication between their laboratories and divisions. In his survey of 55 Japanese and 51 U.S. researchers, Robert Cutler found that Japanese industrial researchers attended more than four times as many technical meetings outside of their laboratories than U.S. industrial researchers. He observed that 80 percent of the Japanese researchers attended technical meetings outside their work location at least twice a month, whereas only 17 percent of the U.S. researchers did this.[9]

One reason for the greater communication between Japanese laboratories and divisions might be the extensive job rotation in

Japanese firms. Mitsubishi rotates its engineers and managers within and between laboratories and divisions. For example, the semiconductor business rotates engineers between its factories to transfer process technology.[10] As described in Part Three, about 50 percent of Mitsubishi's laboratory department managers have had experience as a manager in a factory. It is likely that U.S. laboratories have many fewer department managers with this type of broad experience. Therefore, it is also likely that these U.S. managers have less extensive informal communication networks than managers in Japanese laboratories.

As a result of Mitsubishi's extensive plans and communication networks, project proposals that are submitted to Mitsubishi's R&D selection process are highly integrated. It is very common for multiple departments within a laboratory, multiple laboratories, and different factories to be involved in a single project. For example, Mitsubishi's Fukuoka Works approved funding for 24 separate projects in the fall of 1989. Eight of these projects involved multiple laboratories and ten other projects involved multiple departments within a laboratory. Further, the 24 projects involved a total of 5 different laboratories, 13 different departments and more than 20 different sections within these laboratories. The amount of communication and cooperation required to define and agree on these projects is in itself an impressive accomplishment and a good example of teamwork.

Authority and Responsibilities at the First Review

Up to this point, we have looked at how product development requests, business plans, and informal and formal communication networks help Mitsubishi develop project proposals. Returning to Figure 22-1, all of the project proposals are evaluated in the first review. The general manager and department managers of each laboratory have a great deal of power in this review. Since the laboratories have a finite number of employees (determined by the number of new employees hired by headquarters each year), the laboratories often turn down factory-proposed projects that are not seen as high-growth opportunities for the laboratories and the corporation.[11] The power device department, for example, sometimes had trouble getting projects approved by the laboratories because the laboratories believed that the memory and logic businesses had a much more positive future.[12]

Mitsubishi's corporate headquarters and factories, however, also have some powers that their U.S. counterparts do not ordinarily have. The manner in which these powers overlap provides a good example of shared responsibilities. Since Mitsubishi's corporate headquarters controls the number of college graduates hired each year and since most employees are lifetime employees, corporate headquarters basically controls the number of laboratory employees and the amount of research funding that a laboratory can receive. Personnel from corporate headquarters also evaluate many of the large corporate-funded development projects.

With respect to the works, many of these managers are responsible for evaluating the performance of the laboratory managers. In particular, they evaluate the laboratory personnel according to how well they cooperate with factory personnel and contribute toward the development and application of new technology. Although these responsibilities enable both corporate headquarters and the works to influence the technological direction of the laboratories, this situation also provides the laboratories with a great deal of stability in terms of research funding.[13]

Laboratory personnel are rarely evaluated by their budget performance, since the primary purpose of Mitsubishi's research budgets is to promote the integration of section, department, and laboratory strategies and to promote stability. Instead, research personnel are evaluated on their ability to work with people and their technical qualifications; the number of publications, patents, and conference presentations are important performance criteria. Although some personnel are transferred between sections to reflect changing priorities, labor shortages are primarily made up in the short run through overtime and in the long run through new college graduates.[14]

Contrast this method of balancing power with what exists in U.S. firms. As described in Part Two, many U.S. firms apply "free market" concepts to the development of their research budget. Laboratories typically promote people based on how much funding they are able to collect and how well they balance their budgets. Managers compete with each other for funding from the various corporate, government, and division budgets. Managers who are able to obtain funding are promoted, while those managers who are less successful do not move up.

The problem with this approach is that it does not encourage the integration of research efforts. Instead of being integrators, America's laboratory managers spend a great deal of time balancing their budgets, balancing engineers' schedules, and marketing their research capabilities. More than one manager in a U.S. laboratory has told me that he wished he could "just decide what we're going to do and then do it." Managers spend a great deal of time shuffling research personnel from project to project to balance their budgets. The emphasis on budget balancing in U.S. laboratories is very similar to the emphasis on labor efficiency in many factories. Both of these approaches encourage managers to micro-manage their activities. In the laboratories, the written strategies typically emphasize these budgets; broader, integrating themes are rarely mentioned because these sections and departments are primarily measured by their budget performance.[15]

Authority and Responsibilities in the Final Review

In Mitsubishi's corporate R&D selection process, large R&D projects require additional reviews (see Figure 22-1). The purpose of these reviews is to identify projects that apply to multiple departments in a business or to multiple businesses. Projects that apply to multiple departments can receive funding from business headquarters funds; projects that apply to multiple businesses can receive funds from Mitsubishi's corporate headquarters. Project supporters lobby both business and corporate headquarters personnel to get these funds allocated to their projects.

These reviews are a good example of shared responsibilities and employee developed procedures and plans. Shared responsibilities are involved because decisions are made by people from a number of different organizations. Personnel from the laboratories, the corporate headquarters, and the business headquarters are involved with these decisions. Employee developed procedures and plans are involved because the projects are evaluated using a detailed procedure that was developed and subsequently modified by members of the corporate staff and managers. This procedure uses a fixed set of criteria for evaluating the projects, including a product's expected sales and market share, its technological potential, its stage in the product life cycle, its differentiation from other products, its required technological capabilities, the availability of these capabilities in the laboratories and the corporation, and the applicability of the technology to multiple products.[16]

Based on these criteria, projects are graded into important projects (IP), very important projects (VIP), and strategically important projects (SIP). The categorization is based on the size of the investment, the time to market, and the source of funding. Important projects are funded by laboratory or factory departments and managed by a department manager. Very important projects are funded by corporate or business headquarters and managed by factory or laboratory general managers. Strategically important projects are used to start new businesses or to save strategically important businesses that are having a lot of trouble; they receive corporate funds and are managed at the corporate level.

About 100 proposals reach the final review stage that determines whether a project will be classified as an SIP; Mitsubishi's board of directors makes the final decision. Their decision is heavily influenced by recommendations made by headquarters personnel. Since business groups can receive funding if the project is granted SIP status, they are heavily lobbied by business headquarters personnel.[17]

The sophistication of Mitsubishi's R&D project selection process suggests a strong corporate commitment to R&D; the company's steadily increasing R&D budget confirms this hypothesis. Between 1979 and 1987, Mitsubishi's corporate R&D budget increased from about $300 million to about $800 million, or about 12 percent a year. Because Mitsubishi's growth depends on these internal projects, Mitsubishi's top management spends a large percentage of its time evaluating the VIPs and SIPs that largely constitute Mitsubishi's corporate R&D budget. They spend very little time planning the acquisition and divestiture of businesses, which seems to be the primary activity of America's top management.

Importance of the Informal Communication Network

The informal communication network plays an important part in the identification of potential VIPs and SIPs. These projects typically require the integration of multiple technologies and the identification of appropriate distribution channels. The success of these projects therefore depends on the involvement of various organizations within Mitsubishi. The informal communication network facilitates the identification of the appropriate people in each organization.

Example: The Initial Development of Mitsubishi's Facsimile Machine

A good example of this internal communication network can be found in the initial development of Mitsubishi's facsimile machine. This description summarizes a report written by two Westinghouse engineers, Ed Miller and K. C. Tran, who worked at two different Mitsubishi factories for one year in 1988.[18] In the mid-1970s, Mitsubishi's television-producing factory, the Kyoto Works, was faced with surplus capacity due to the maturation of the television market. Many people at the Kyoto Works were searching for new products that could be manufactured there. One such person was Mr. Matsumara, who was then manager of the video manufacturing department and was very interested in the facsimile machine. Since he lacked the necessary technical information concerning this product, he decided to discuss the topic informally with an old friend from Kyoto Works, Mr. Washio, who had since become a department manager at the central research laboratory.

This conversation triggered a series of conversations that quickly led to the submission of a proposal to Mitsubishi's formal R&D selection process. Since Washio was also unfamiliar with the essential technologies for the facsimile machine, he contacted another former colleague, Mr. Uetake, who worked in another Mitsubishi laboratory. Uetake spoke with several people, one of whom, Dr. Ryoichi Onishi, had considerable expertise in signal coding and expansion technology. This technology was critical to the development of the facsimile machine and became Mitsubishi's technological advantage in the facsimile business. Based on his conversation with Onishi, Uetake was convinced that Mitsubishi should develop its own facsimile machine. Therefore, he contacted his boss, the general manager of the laboratory, to begin developing a group of people who could submit a formal proposal to Mitsubishi's R&D selection process.

Once Uetake had spoken with the laboratory's general manager, the facsimile project became part of Mitsubishi's formal R&D selection process, and there was an accepted procedure for the general manager to follow. The general manager contacted the director of Mitsubishi's electronics business group, Mr. Yoshida, who acted as the spokesperson for the facsimile project. Through a series of formal meetings between various laboratory representatives, the Kyoto Works, and the headquarters business sales organization, a project team was developed consisting of members from the Kyoto Works, the central research laboratory, the consumer products laboratory, the headquarters business sales group and the electronics business group. The project team developed a proposal, which passed through each of the reviews in the corporate R&D selection process. The development of the facsimile was classified as an SIP, and corporate funds were allocated to it. Mitsubishi has subsequently become one of the leading suppliers of facsimile machines.

Mitsubishi's informal communication network was a key reason for the success of this project. Without this informal network, the technology and personnel needed to develop the facsimile machine might never have been identified. This network was developed through numerous corporate projects and Mitsubishi's emphasis on job rotation. Many of the early conversations concerning the facsimile machine occurred between engineers who had at one time worked together but who subsequently had been transferred to other works or laboratories.

Defining the Technological Needs of Semiconductor Equipment: Teamwork Between a Business and Its Customers

THIS CHAPTER describes how Mitsubishi's semiconductor business defines the technological needs of its semiconductor equipment. Although several organizations within Mitsubishi are involved with these activities year-round, many of these technological decisions are made just before the corporate R&D selection process begins in September. The organizations primarily involved in defining the technological needs of Mitsubishi's semiconductor equipment are the semiconductor equipment department, the semiconductor factories, the headquarters of Mitsubishi's semiconductor business, and Mitsubishi's large scale integration (LSI) laboratory. The way in which the various organizations define these technological needs is a good example of how a Japanese firm manages the linkages between a laboratory, division headquarters, factories, and their suppliers.

The various Mitsubishi organizations primarily use a strategy of incremental improvement to define the technological needs for the semiconductor equipment. As outlined in Figure 23-1, each unit of technology is developed and tested; an equipment technology plan is updated based on these tests and on existing production problems. The new unit is incorporated into the next version of equipment; it is tested and the plan is updated again. A feedback loop between equipment test and unit test enables the recently incorporated unit or the next unit developed to be modified according to the most recent tests. As mentioned earlier, the emphasis on incremental improvement of products actually represents a highly evolved technology development process in that technology *flows* from laboratories to divisions as opposed to moving in large chunks. This flow of technology requires close feedback and teamwork between the various organizations involved with defining the technological needs of Mitsubishi's semiconductor equipment.

This situation is not unique to Mitsubishi's semiconductor business. Howell Hammond, a director of research for Kodak in Japan, found that Japanese engineers use a process similar to that shown in Figure 23-1, in which they continually repeat cycles of idea, prototype, and test.[1] Another study found that many of Japan's semiconductor factories have close relations with their equipment suppliers and emphasize incremental improvements in the semiconductor process and equipment.[2] Several other observers have found that other Japanese firms emphasize the continual improvement of products[3] and that Japanese firms are typically vertically and horizontally integrated.[4] Several of these same people observe that Japanese firms closely coordinate the activities in these businesses, particularly in the electronics industries.[5]

Defining the Technological Needs of Packaging Equipment

Packaging equipment assembles an individual chip into a package; it includes die bonding, wire bonding, molding, and lead process equipment. Mitsubishi's semiconductor equipment department, semiconductor chip factories, and the headquarters

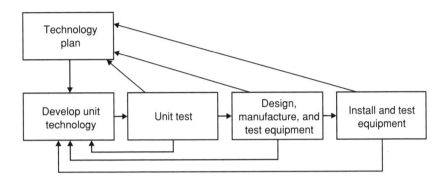

Figure 23-1. Incremental Improvements in Technology and Equipment

of its semiconductor chip business (IC assembly engineering department) are involved with defining the technological needs of Mitsubishi's packaging equipment. These organizations work together in a manner similar to that described in Chapter 12.

Responsibility for defining the technological needs of Mitsubishi's packaging equipment is shared between the SED and these other organizations; a formal communication network, the assembly technology communication meetings, exists to support these shared responsibilities. These meetings are attended by representatives from each of Mitsubishi's semiconductor factories, each appropriate equipment group in the SED, and each appropriate equipment group in the IC assembly engineering department.

Technology Plans

Technology plans, a type of employee developed plan, are an important part of the department's incremental improvement philosophy. They describe how each unit of technology is to be improved and integrated with the equipment. For example, Figure 23-2 shows an example of such a plan for wire bonder equipment. A wire bonder attaches gold wires from each of an individual chip's pads to each of an individual package's metal leads. Its most important components are shown on the left side of the figure. They include a bonding head, vision system, key units (torch assembly and ultrasonic power), and a feeder system. The evolution of these units is shown from left to right over a three-year period. Some of the problems with two of the wire bonder's key components (bonding head and vision system) and the planned solutions to these problems are also summarized in this figure.

The purpose of these technology plans and unit tests is to improve the equipment as quickly as possible, adding new technology only when it is ready. The technology plans identify the critical technologies and the time frames in which they will be developed and implemented. They enable the department to incrementally improve the equipment using parallel developments of multiple technologies while simultaneously developing the technology needed for large improvements in equipment performance. The unit tests enable the department to introduce new technology only after it has been tested and verified.

The department's product-oriented organization enables each of the department's seven equipment groups to be primarily responsible for writing their own plans and testing each unit of technology. In addition, since the semiconductor chip business requests many of the technology improvements and participates in the tests, responsibility for these activities is also shared with the headquarters of Mitsubishi's semiconductor business and with the semiconductor factories. Many of these decisions are made in the assembly technology communication meetings. In addition, a summarized version of these assembly technology plans becomes part of the one-, three-, and five-year technology plans developed by Mitsubishi's semiconductor business.

Other researchers have observed similar types of plans in use in other Japanese firms. Sheridan Tatsuno, for example, describes several of these plans, including technology road maps

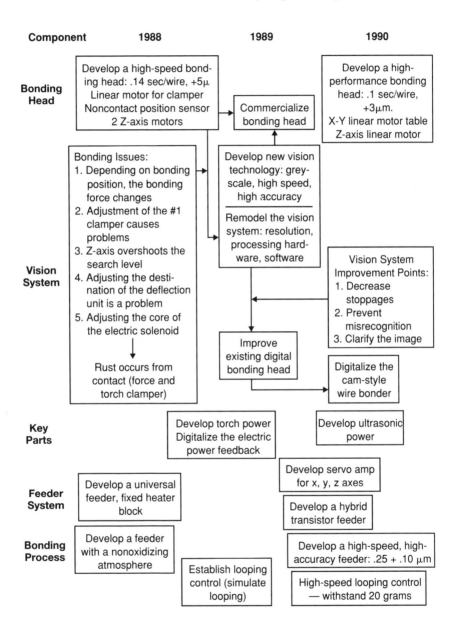

Figure 23-2. Example of a Technology Development Plan for Wire Bonder Equipment

(similar to Figure 23-2) and technology trees that show how a firm's fundamental technologies grow into technologies specific to a particular product.[6] Richard Rosenbloom and Michael Cusumano, in their study of the birth of VCRs, found that Sony and Matsushita focused many incremental improvements on customer needs through "strategic consistency;"[7] this sounds a lot like detailed technology plans. Yokogawa Electric uses technology trees to identify the relationship between basic research and the resulting products. A number of successful U.S. firms also use these technology plans or maps.[8]

Intersecting the Elements of New Equipment Technology

Mitsubishi uses these plans to implement a strategy it calls "intersecting the elements of new equipment technology." It was one of the SED's four main strategies in its 1989 strategic plan, and almost every Mitsubishi factory I visited in 1989 was also familiar with the concept. This strategy uses the technology development plans to incrementally improve each unit of technology and the equipment that uses these unit technologies. Along with employee developed procedures and plans and the use of standard forms, this strategy enables Mitsubishi to manage the flow of product technology as opposed to only managing each individual product development project. By managing the flow of product technology, Mitsubishi is able to incrementally improve its semiconductor equipment and other products very quickly.

Employee developed procedures and plans and standard forms enable the SED to design new equipment quickly without a great deal of managerial involvement. The department's well-defined design methodology is a good example of employee developed procedures. Since this design method is used by all of the engineers in the department, it basically defines the activities needed to design a new equipment model in the minimum amount of time. Standard forms are used to develop the detailed schedules for designing a new equipment model. Using these forms, a senior member of the equipment group develops a macro-level schedule to incorporate one or more new units of technology into an existing equipment type. The junior members in each group then develop individual schedules, using other standard forms.

A combination of these standard forms, the SED's well-defined design methodology, and the technology plans (i.e., intersecting the elements of new equipment technology) enable the department to focus on the flow of technology rather than on individual projects. Since the design methodology requires almost no involvement of the section managers and only some involvement of the equipment group leaders and the assistant group leaders, these managers can focus on the technology development plans and on whether these new technologies will improve semiconductor equipment performance. Through the assembly technology communication meetings, these people work with the semiconductor factories and the IC assembly engineering department to determine which technologies will improve equipment performance. New technologies are often developed in parallel, with as many as four levels sometimes developed at the same time.

Defining the Technological Needs of New Wafer Fabrication Equipment

Mitsubishi's semiconductor equipment department and its semiconductor chip business also work together to define the technological needs of its wafer fabrication equipment. Wafer fabrication equipment produces the wafers that contain the individual chips; it includes such equipment as plasma etching and chemical vapor deposition equipment. As with the packaging equipment, shared responsibilities and a formal communication network play a strong role in the development of this equipment. Responsibility for defining the technological needs of the wafer fabrication equipment is shared between the SED and the Mitsubishi semiconductor business; a formal communication network (the monthly wafer process technology communication meetings) exists to support these shared responsibilities.

Until early 1989, the members of the wafer process technology communication meetings were similar to the members of the assembly technology communication meetings. Representatives from each of Mitsubishi's semiconductor factories, each of the appropriate equipment groups in the SED, and each of the appropriate groups in the wafer process engineering department attended the wafer process technology communication meetings. The only difference between the wafer process technology and assembly technology communication meetings was the actual equipment groups represented in these meetings (wafer fabrication versus packaging equipment).

Due to a change in the Mitsubishi semiconductor business strategy, several of Mitsubishi's laboratories began attending the wafer process technology communication meetings. This is a good example of how a change in business strategy dictates a change in the formal communication networks and the way in which responsibilities are shared between various parts of a Japanese firm.

A Change in Business Strategy Dictates a Change in Teamwork

The shift in business strategy concerned the source of wafer fabrication equipment for the development of new memory chips. As discussed in Part Two (Chapter 7), new memory chips require the development of new wafer fabrication processes, which in turn require the development of new wafer fabrication equipment. The speed with which new equipment is developed often determines the speed with which new memory chips can be developed — and the firm that first produces a new chip often ends up being the most profitable.

New generations of memory chips have been developed in Mitsubishi's LSI laboratory for a number of years. Although the SED supplied the LSI laboratory and Mitsubishi's semiconductor factories with some of its wafer fabrication equipment for a few of those years, the laboratory purchased much of its most advanced equipment from a variety of non-Mitsubishi sources. In the late 1980s, however, Mitsubishi decided that to compete in the memory and logic chip businesses, it needed a better source of wafer fabrication equipment. It felt that outside suppliers of wafer fabrication equipment were not responding quickly enough to its requests and were unwilling to adequately customize the equipment to Mitsubishi's production needs.[9] The company felt the need for an internal capability that could respond directly to its needs, particularly in the area of etching and deposition equipment.

Members of Mitsubishi's semiconductor business proposed a five-year project (the Sub-micron Development Project) to develop four generations of etching and deposition equipment. Together these four generations of equipment would enable Mitsubishi's semiconductor business to reduce the line widths on its chips from about one micron to much less than .5 microns. This equipment would be jointly developed by the SED, the headquarters of Mitsubishi's semiconductor business (wafer process engineering department) and several of Mitsubishi's laboratories. The project proposal was submitted to Mitsubishi's corporate R&D selection process; it was eventually funded as a VIP and received funding from the headquarters of the semiconductor business.

The magnitude of the Sub-micron Development Project required a significant change in the relationship between several organizations within Mitsubishi. This was reflected in the written strategies of the SED and the relevant equipment section.

Both emphasized the need for the department and the section to develop closer ties with the LSI and other laboratories in this project. The monthly wafer process technology communication meetings were also modified to facilitate the development of this equipment. The new meeting included representatives from the SED, the headquarters of Mitsubishi's semiconductor business, the LSI laboratory, the manufacturing development laboratory, and the central research laboratory.[10]

In the fall of 1989, the general manager of the LSI laboratory visited the SED to formalize the change in relationship between the laboratory and the department. He gave a one-hour presentation to all the members of the department and listened to presentations by several department engineers concerning their work in the Sub-micron Development Project. He discussed the history of semiconductor production in Mitsubishi, important technological trends, and Mitsubishi's strategy for dealing with these trends.[11] This presentation is a good example of the emphasis on oral communication at Mitsubishi. It was excellent for morale and provided the members of the SED with an overview of the company's semiconductor business strategy.

Incremental Improvement of Wafer Fabrication Equipment

The Sub-micron Development Project is also a good example of incremental improvements. The project defined the development of four generations of this equipment and several incremental improvements in unit technology for each generation. Each generation of equipment would be installed and tested in the LSI laboratory.

Employee developed procedures and plans were an important part of these incremental improvements. An equipment plan was developed for each of the four generations of etching and deposition equipment. These plans are very similar to the plans shown earlier in this chapter for the wire bonder equipment (see Figure 23-2). Individual units of technology were to be developed by the SED and the various laboratories involved in the VIP and intersected and commercialized in a manner similar to the wire bonder equipment. Therefore, more than four prototypes would exist for both the etching and deposition equipment, since technology was to be added incrementally to each type of equipment. As with the wire bonder equipment, senior members of the etching and deposition equipment groups managed this "intersection of unit technologies." The senior members of these groups developed the macro-level schedules while the junior members developed and managed the individual schedules.

The etching and deposition equipment also depended on another employee developed procedure and plan called a *technology map*. The various organizations involved with the Sub-micron Development Project used these maps to develop the technology plan for the project. These maps portray on multiple axes the full range of potential operating conditions for a specific product, in this case for both etching and deposition equipment. Existing equipment is shown as boxes on these maps. By knowing the performance capabilities for each of these equipment types, it is easier to identify the appropriate characteristics for the next equipment model.

For example, the key operating variables for plasma etching equipment are pressure, energy, and reactant type. The key performance characteristics are the anisotropic etching rate, the ratio between the isotropic and the anisotropic etching rate, the uniformity of the etching rate across a wafer, and the ratio between the photoresist etching rate and the removed material's etching rate. Mitsubishi's technology maps showed the existing etching equipment on these maps in terms of pressure, energy, and reactant type. Experiments had been performed on the existing equipment to determine the performance characteristics over the potential range of operating characteristics. Mitsubishi's semiconductor business used these technology maps to develop the sub-micron technology plan, the individual equipment plan, and the experimental plans. As new equipment is developed, experiments are performed; new data are entered on the technology map, and the equipment and experimental plans are modified accordingly.[12]

Shared responsibilities and formal communication networks complement the use of these technology maps. These maps are used as a form of communication between the various organizations involved in a formal communication network (in this case, the wafer process technology communication meetings). These organizations use the maps to develop technology plans for the deposition and etching equipment. In addition, within Mitsubishi's semiconductor business, section managers, department managers, and sometimes general managers use these maps to develop long-term plans and to integrate these plans with those developed by other sections, departments, factories, and laboratories within the business. For example, the SED's department manager spends a great deal of his time discussing trends in the semiconductor industry with other high-level members of Mitsubishi's semiconductor business. He uses these discussions to integrate the SED's strategies and plans with other parts of the semiconductor business and to communicate these strategies and plans to his department.[13]

The Benefits of Modular Design

Another factor that helps Mitsubishi define the technological needs of its semiconductor equipment is the use of modular design. As discussed in Part Five, modular design is an important part of the SED's design strategy (an example was described for etching and deposition equipment in Chapter 17). The department originally planned to use a similar equipment structure for all four generations of both the etching and deposition equipment. To reduce development and manufacturing costs, only the two processing chambers would be different for the etching and deposition equipment. The input and output chambers would be the same for each equipment type.[14]

The use of a modular design enables Mitsubishi to focus its technological improvements in the etching equipment on only one of its four chambers. Only one chamber has to be redesigned as the four generations of etching equipment are developed. The modular design also enables the department to apply the same technology to both the etching and deposition equipment. Since two of the four chambers are the same for each equipment type, only two of the chambers require specific development for one equipment type or the other.

This modular design also helps the engineers from the etching and deposition equipment groups work together. These two types of equipment are very dependent on each other; one deposits the material (i.e, chemical vapor deposition equipment) and the other etches it (i.e., etching equipment). Both types of equipment must be able to process the same type of materials; any change in one equipment's design will affect the design of the other. By using common elements of technology (in this case the input and output chambers) these two equipment groups will stay closely in touch with each other.

———

Developing the Elements of New Equipment Technology: Teamwork Among Laboratories, a Business, and Suppliers

M ITSUBISHI'S semiconductor equipment department also works closely with the company's laboratories to develop elements of new equipment technology. This chapter discusses how one particular equipment group within the SED, the wire bonder equipment group, worked with various laboratories to develop two elements of technology for the wire bonder equipment: a bonding head and the vision system.

Example: Incremental Improvement of a Digital Bonding Head

A bonding head feeds wire to the bonding point, forms a ball on the end of the wire via a high-voltage current, and bonds the wire ball to the chip pad and lead frame using an ultrasonic force. The development and improvement of a digital bonding head is a good example of incremental improvements using the technology plans. Several improved versions have been introduced, since a digital bonding head was first used in a Mitsubishi factory in March 1988. As Figure 24-1 shows, four different bonding heads were simultaneously under development in 1989.

Three of these bonding heads were incremental improvements to the existing bonding head; the fourth represented a new generation of bonding equipment called tape-automated bonding. The incremental improvements are part of the activities included in the box entitled "commercialize bonding head" at the top of Figure 23-2. The goals for the new generation of bonding equipment are shown in the box at the top right of the figure.

The development of the first digital bonding head began in 1984. In 1987, it was transferred to the SED for testing, modification, and integration with the other elements of the wire bonder equipment. Before it was transferred, however, the next digital bonding head was being planned in the assembly tech-

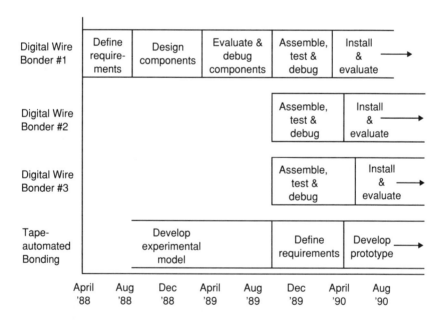

Figure 24-1. Incremental Improvement and Parallel Development of New Bonding Heads

nology communication meetings. In these meetings, the SED, the semiconductor factories, and the IC assembly engineering department decided that the manufacturing development laboratory should begin developing a higher speed digital bonding head as soon as the first digital bonding head was being installed in Mitsubishi's semiconductor factories. The project was approved in the fall of 1987 in the corporate R&D selection process. Work began in the spring of 1988, just as the first factory installations of the equipment containing the first digital bonding head were occurring.[1]

Development of the Second Digital Bonding Head

The second digital bonding head is a good example of shared responsibilities and formal communication networks. In the first four months of the project, the laboratory worked with the wire bonder equipment group to define the equipment's specifications and determine the exact responsibilities and task breakdowns for the project. The specifications and responsibilities were documented in August 1988. The laboratory was to develop the bonding head and the x-y table, whereas the wire bonder equipment group was to develop the vision system, frame feeder, control system, and the equipment's enclosure. After developing these individual elements of technology, the laboratory and this equipment group shared responsibility for integrating the elements of technology and evaluating the equipment.[2]

According to the wire bonder equipment group's assistant group leader, Kazuhiro Kawabata, this situation is common in Mitsubishi. Instead of having a clear handoff between the laboratories and the works, there is a gradual transfer of technology and responsibility for the technology from the laboratories to the works. Mr. Kawabata claims that Mitsubishi's divisions do not expect the laboratories to deliver a product that is ready to be sold on the market or used in an internal production environment. The divisions expect the laboratories only to develop a technology that the divisions lack the necessary resources to develop themselves. The laboratories are expected to deliver this technology to the factories before all of the bugs have been worked out. After the physical technology is delivered to the works, the laboratory collaborates with the factories to improve the technology incrementally.[3]

During the development of the second digital bonding head, there was constant communication between the laboratory and the wire bonder equipment group. According to one engineer on the laboratory's development team, he met with members of this group about once a month and talked with these people about twice a month on the telephone.[4] Equipment users were given status updates via the assembly technology communication meetings, which occurred about once a month. Another manager in a Mitsubishi laboratory confirms that this type of communication is very common. He claims that laboratory and factory engineers typically meet several times and talk on the phone about twice a month during the execution phase of a project.[5]

Employee developed procedures and plans supported further incremental improvements in the second version of the digital bonding head. The responsibilities and schedules for this project were described in several detailed plans that had been developed by engineers from the wire bonder equipment group and the manufacturing development laboratory. When the second version of the digital bonding head was transferred to the wire bonder equipment group, a one-year plan to commercialize the bonding head's technology in three phases had already been developed (see Figure 24-1).

The one-year plan depended on several other technology plans developed by the employees. Figure 24-2 represents the way in which the wire bonder equipment group continually documents the bonding head's needed improvements, the technologies needed to improve these points, the specific improvement projects, and their interdependencies. This particular plan is merely a subset of the equipment technology plan described in Chapter 23 (see Figure 23-1). The equipment technology plan describes how the elements of technology evolve and are integrated. Figure 24-2 describes the needed improvements, the technologies needed to make these improvements, and the specific improvement projects for one technological element of the wire bonder, in this case the bonding head.

All of these plans — the one-year plan to commercialize the digital bonding head and the plans shown in Figures 23-1, 23-2, and 24-2 — depend on input from the manufacturing development laboratory, semiconductor factories and the IC assembly engineering department. In the case of the one-year plan, it was based on the laboratory's test results and constant evaluations of the existing bonding head's performance in Mitsubishi's semiconductor factories. The wire bonder equipment group had worked closely with the laboratory during the equipment's testing; it knew which elements of the new bonding head were ready for commercial production and which elements required further testing. Although much of this communication occurred informally, the consensus needed to develop the technology plans depended heavily on the formal communication networks that connect these organizations. These networks include the assembly technology communication meetings and the transfer of engineers between the laboratories and the wire bonder equipment group.[6]

Incremental Improvements to the Second Digital Bonding Head

The wire bonder equipment group's one-year plan to commercialize the new digital bonding in three phases is a good example of how Japanese firms incrementally improve a product. It is also a good example of the benefits from incremental product improvement and the importance of Japan's corporate culture in supporting this improvement.

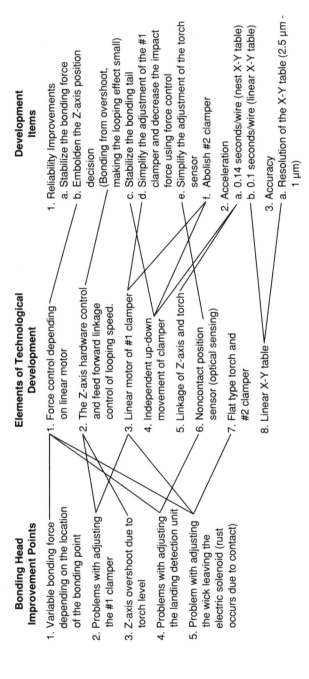

Figure 24-2. Improvement and Development Items for the Bonding Head System

In the first phase of the plan, a position sensor that depended on contact between two surfaces was replaced with an optical sensor because the contact sensor had corroded easily and required excessive maintenance. This model was delivered in early 1989. New wire clampers that used linear motors were added in the second phase of the project; this model was first delivered in August 1989.[7] In phase three, two Z-axis motors replaced one Z-axis motor so that certain movements by the bonding head could be performed simultaneously. The new Z-axis motors reduced the bonding cycle time from 0.18 seconds to 0.15 seconds. A bonding head containing the two Z-axis motors was to be delivered by April 1990.[8]

The three-phase plan provided Mitsubishi with a number of benefits. First, it enabled the wire bonder equipment group to test a number of small improvements individually. Numerous tests were performed on the optical sensor, wire clampers, and Z-axis motors before they were integrated with the rest of the bonding head and wire bonding equipment. These tests are a good example of the utility of multifunctional employees. The tests were performed by the regular design engineers rather than by a special "test" engineer, as is typically the case in U.S. firms.[9] The department does not employ any special test engineers; most of the design engineers are multifunctional, performing both the design and the test functions.

The second benefit from using this three-step plan is that Mitsubishi's semiconductor factories could start taking advantage of the wire bonder's new capabilities much sooner than if the technology was commercialized in one step. Instead of waiting for the two Z-axis motors to be developed, the factories were using the new optical sensors one year and the wire clampers six months before the new motors had been completely integrated with the new bonding head.

These incremental improvements also depended a great deal on teamwork within the wire bonder equipment group and on a formal communication network, the regular weekly wire bonder equipment group meetings. For example, although the Z-axis motors were primarily developed by two engineers in that group, development also depended on input from other members of the group. Masayuki Hiroki, a junior software engineer, developed the software and Yasufumi Nakasu, a junior electrical engineer performed the mechanical tests on these motors. In each weekly wire bonder equipment group meeting, these two engineers summarized the software problems and their proposed solutions. Since other members of the group have several years of experience with the wire bonder equipment, they were able to provide these young engineers with valuable advice and to determine if other engineers needed to be involved with the development of the two motors. If help from other engineers was needed, smaller meetings were held after the regular weekly meeting.

The Next Generation of Bonding Heads

Long before the second digital bonding head was transferred from the laboratory to the wire bonder equipment group, plans for the next generation of bonding technologies were being discussed in the assembly technology communication meetings. Until early 1989, the next digital bonding head was expected to be similar to the last two generations of digital bonding heads, only faster. However, a new bonding technology called tape-automated bonding had been under development by the manufacturing development laboratory and the IC assembly technology department for several years. The wire bonder equipment group was kept informed of those activities in the assembly technology communication meetings. Due to the increasing number of leads per package, it became clear to many of the participants in these meetings that the conventional bonding technology might not be able to adequately handle the increasing size of memory and other chips. In 1989 it was decided in the assembly technology communication meetings that the wire bonder equipment group should begin developing a tape-automated bonding machine.

During the summer of 1989, the wire bonder equipment group developed an experimental machine, primarily using parts from the existing wire bonder and die bonder. And in the fall of 1989, this group began funding the manufacturing development laboratory to help merge the laboratory technology with this group's technology and the die bonder equipment group's technology. As usual, the first step of the project was to transfer one engineer temporarily between the two organizations to define the specifications and schedule for the project.

In this case, an engineer from the manufacturing development laboratory, Mr. Hokamura, spent one month in the SED performing experiments on the wire bonder's preliminary unit and studying the technology of the wire bonder and die bonder equipment. Hokamura compared several versions of the wire bonder equipment group's experimental version with each other and with the manufacturing development laboratory's technology.[10] This is a good example of job rotation. Although Hokamura was not officially transferred to the SED, he participated in the wire bonder equipment group's regular activities during his month with them. He attended its daily standup meetings and its weekly equipment group meetings.

Example: Incremental Improvements to the Vision System

Japan's corporate culture also helped the SED incrementally improve another element of equipment technology: the wire bonder's vision system. The wire bonder equipment uses a vision system to locate the chip's pads before the bonding head attaches gold wires from these pads to the leads of the chip's package. Like the bonding head, the vision system is also incrementally improved, with several of these improvements being developed in parallel. As Figure 24-3 shows, five vision systems were simultaneously under development in 1989. Four improvements were incremental advances to the existing system; the fifth represented a new generation of vision systems, a grey-scale vision system. These improvements were supported by the wire bonder equipment's technology plans. The five improvements are part of the box directly below the "commercialize bonding head" box in Figure 23-2. (Most of the other boxes shown in Figure 23-2 also have schedules similar to the ones shown in Figures 24-1 and 24-3.)

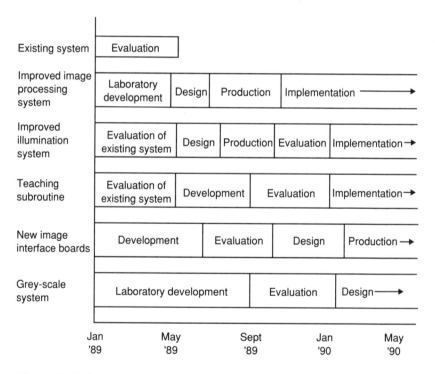

Figure 24-3. Incremental Improvement in and Parallel Development of New Vision Technology

The incremental improvements also depend on several other aspects of Japan's corporate culture. Detailed plans are developed by the employees for these improvements in the vision system in a manner similar to the bonding head improvements. For example, the wire bonder equipment group used the matrix shown in Figure 24-4 to relate customer needs with the vision technology. Responsibility for developing and implementing this type of plan is shared between this group, equipment users, and the appropriate laboratories. A formal communication network supports these shared responsibilities, in the form of the assembly technology communication meetings, as well as several other meetings and employee transfers between the wire bonder equipment group and the appropriate laboratories.

In 1988, it was decided in the assembly technology communication meetings that the existing vision system needed several improvements that were eventually added in five individual projects. The SED, the semiconductor factories, and the IC assembly engineering group decided that the wire bonder equipment group should work with the industrial systems laboratory to make some of these improvements to the existing vision system. Since the industrial systems laboratory had developed the existing vision system a few years before, it was the logical laboratory to work with this group.

Teamwork in the Laboratory: Developing the Conceptual Design

Two members from the wire bonder equipment group, Kazuhiro Kawabata and I, were sent to the industrial systems laboratory in the spring of 1989 to develop a conceptual design for the new vision system. We discussed the existing system, the costs of alternative vision systems, and various experimental data with two key members of the vision group, Kazuhiko Sumi and Shinjiro Kawato. The wire bonder equipment group had collected this experimental data to better understand the problems with the existing system.

Improvement Objectives / Elements of Technology	Improvement Points for Existing System				Development Items				Level of Priority
	Eradicate production of bad chips	Decrease frequency of equipment stoppages	Clarify the images	Improve the accuracy	Increase the speed	. . .	. . .		
Appropriate illumination (select type of light source)	◯	◯	◎	◯					1
Remodel the lens: Improve the magnification, focus	◯	◯	◎	◎					5
Select the optimum camera, Improve reduction in image size	◯	◯	◯	◯					4
Improve the H/W processing (improve the counter, % reduction)	◎	◎							3
Automate the setting of choice criteria: threshold level, score limit, target pattern	◯	◯		◯					6
Automate the setting of the offset				◯					7
Speed up the recognition of individual leads					◎				9
S/W processing — Utilize four independent references	◎	◎							2
Software processing — Improve the algorithm	◎	◎							8

◎ = Large effect   ◯ = Effect

Figure 24-4. Vision System Development and Improvement Items

However, the most interesting aspect of the visit was the way in which additional data was collected during our visit to the laboratory. To gain a better understanding of the advantages and disadvantages of alternative vision systems, the majority of the one-week visit was spent collecting more experimental data. This experimental data was then used to further refine the conceptual design for the vision system. After receiving suggestions about which equipment might be useful and instructions on how to use the image processing computers, Kawabata and I tested a variety of illumination and lens systems using the semiconductor chips that had been found to have the largest yield problems. Data were collected on these experimental systems and compared to data previously collected for the existing vision system.

The fact that Kawabata and I performed these experiments ourselves is an interesting example of the use of multifunctional employees and shared responsibility. We were performing an activity that would probably be done by a laboratory specialist in the United States. According to the vision technology group's leader, Mr. Kawato, visiting engineers often perform their own experiments when they visit the laboratories since the engineers know more about their own needs than do the laboratory personnel.[11]

The strategy makes sense. If the laboratory personnel perform the experiments, they must write a report, the factory responds to the report, and typically more experiments are performed. In the long run, it is probably cheaper and faster for visiting engineers to perform the experiments themselves. However, this situation is probably very rare in U.S. firms, because it requires a laboratory and factory to have a long-term relationship; both must be very flexible in the way they define work. Short-term or arm's-length relationships and detailed job descriptions prevent many U.S. firms from building this type of relationship between their factories and their laboratories.

Teamwork at the Works: Developing a Detailed Action Plan

After Kazuhiro Kawabata and I returned from the laboratory visit, the wire bonder equipment group developed a six-month schedule to improve incrementally the existing vision system in three improvement projects. The three projects were defined so that each could be developed in parallel and could be added incrementally to the existing vision system without the other projects. This enabled each project to be tested individually, gave the wire bonder equipment group quick feedback on each improvement project, and allowed Mitsubishi's semiconductor factories to begin benefiting from the new technology before all of the projects were completed.

The planning and implementation of the three projects required a great deal of teamwork between certain members of the wire bonder equipment group. Although the projects were defined so that projects could be added independently of each other, the achievement of this goal required constant feedback between the engineers in the group. In the first project, a mechanical engineer, Toshiyuki Egashita, was assigned to design a new illumination system for the vision system. In the second project, an electrical engineer, Naoki Miyazaki, was assigned to design new image processing hardware, and one of the software engineers, Naomi Yokoyama, was assigned to design the new image processing software. In the third project, I was assigned to develop a software package that could automatically "teach" the vision system several aspects about a new chip model.

Several elements of Japan's corporate culture supported this teamwork and helped the group develop and implement its six-month schedule. Due to the product-oriented organization, most of the engineers involved with the three projects worked in the wire bonder equipment group. These engineers developed detailed plans and schedules for these projects. Senior members of the group developed macro-level schedules while the other members of the group developed individual schedules based on the macro-level schedules. These schedules were developed using standard forms. In the open office, these engineers sat next to each other and they summarized their activities on these sub-projects in the daily standup meetings.

Teamwork with a Laboratory and Supplier: Design of the Hardware

The industrial systems laboratory and a supplier of printed circuit boards were both closely involved with several of these improvement projects. One of the three projects, the design of the electrical hardware, is the best example of teamwork between this laboratory, the wire bonder equipment group, and its supplier. Shared responsibilities and formal communication networks were particularly important in this project. Responsibilities were shared between the wire bonder equipment group and the other two organizations; two formal communication networks were established to support these shared responsibilities.

First, I'll discuss the relationship between the wire bonder equipment group and the industrial systems laboratory. As described above, Kazuhiro Kawabata and I had already discussed the conceptual design of the image processing algorithm with the laboratory. However, instead of having the laboratory write a report or develop the detailed design at this point, the wire bonder equipment group sent one of its electrical engineers, Naoki Miyazaki, to the laboratory for a month to design the hardware under the tutelage of the laboratory's expert in this particular vision technology, Kazuhiko Sumi.

Miyazaki spent the month of June at the industrial systems laboratory studying documents and discussing the detailed design of the system with Kazuhiko Sumi. They met about three hours per day for half of the days and about one hour per day for the remainder. After Miyazaki returned to the SED, the engineers continued to communicate once a week for a couple of months via facsimile or telephone.[12]

Within two weeks of his return to the SED, Miyazaki began working with the wire bonder equipment group's printed circuit board supplier. This supplier does the board layout and manufactures most of the boards for the wire bonder equipment group and the other equipment groups in the SED. An electrical engineer from the supplier, Mr. Utsumi, moved into an apartment about a mile from the SED and began coming to the department every day as a quasi-departmental employee. He was given a desk in the wire bonder equipment group's area of the office and attended the group's daily meetings and weekly equipment group meetings.

Miyazaki and Utsumi worked closely together on this project. They met about four hours per day in the first month of Miyazaki's relocation to the wire bonder equipment group. Their interactions gradually decreased to about one hour a day in the succeeding months as the individual tasks became better defined.[13] The logic design and layout were completed for two printed circuit boards by the end of August, and the first manufactured printed circuit boards were received in November.

After the detailed design was completed, Utsumi began studying other aspects of the wire bonder as part of a training program. He studied under the tutelage of Miyazaki to better understand the wire bonder equipment group's needs. He was still working in the wire bonder equipment group when I left in December and he had become involved with several other projects associated with the wire bonder equipment.

*Teamwork Between Different Equipment Types:
Further Improvements to the Vision System*

While Miyazaki and Utsumi were designing the new hardware for the wire bonder equipment's vision system, four other improvement projects were proceeding in parallel with the design of the new hardware. Three of these projects were defined so that they could be integrated with the existing vision system independent of each other; the fourth project involved the development of a new generation of vision systems. Since two of these projects involved multiple applications of the same technology, they are an interesting example of teamwork between multiple equipment groups. One project was the development of new image interface printed circuit boards; the other was the development of a grey-scale vision system, a new generation of vision technology.

New image interface boards were originally developed for another type of assembly equipment called a die bonder. Portions of this project were described in Part Five as an example of a manufacturing cost analysis. Here, however, this project is described as an example of applying one assembly equipment's technology (die bonder) to another type of equipment, a wire bonder.

After Miyazaki had finished the detailed design of the image processing system's hardware, he began investigating the image interface boards. He began his investigation by analyzing the image interface boards that had been developed for the die bonder equipment. He performed tests on the boards, read the detailed documentation, and spoke with the engineers who developed the boards.[14]

Several elements of Japan's corporate culture helped Miyazaki quickly apply aspects of this technology to the wire bonder equipment. The design documentation was readily accessible since it was stored in the public files. The open office made it easy for Miyazaki to question Shoyo Hayashi, who had originally designed the image interface boards.

Another of the vision system's improvement projects involved multiple equipment types. Mitsubishi's IC assembly and IC wafer process engineering departments had been funding the development of a general purpose grey-scale vision system for multiple types of semiconductor equipment, including the wire bonder equipment. The project's progress was discussed in the assembly technology communication meetings, which included representatives from the wire bonder equipment group and other organizations. In the fall of 1989, these organizations decided that the wire bonder equipment group should fund the industrial systems laboratory to apply this technology specifically to the wire bonder and develop a prototype. The funding was approved during the fall, and work began in April 1990.

This formal communication network (the assembly technology communication meetings) enabled Mitsubishi's semiconductor business to apply generic technology to a specific application (the wire bonder equipment). Since new vision technology is very expensive, multiple applications are needed to make its development economical. This particular formal communication network enables multiple equipment types to develop and apply generic technology.

Another formal communication network helps the SED apply generic vision technology to multiple equipment types. At about the same time that the assembly technology communication meetings were deciding that the wire bonder equipment group should develop a grey-scale vision system, the SED initiated a working group to apply this vision technology to multiple equipment types. This working group, which included a representative from each equipment group, is similar to the departmental working groups that deal with the use of standardized software, hardware, and mechanical technologies in various types of semiconductor equipment. Due to the increasing importance of vision technology to each type of semiconductor equipment, the SED decided to use the vision working group in a manner similar to these other working groups. The vision technology working group was made responsible for comparing the vision needs of each equipment type and defining the vision technology commonly needed in multiple equipment types. This working group will make the application of vision technology more economical to the semiconductor equipment department.

Developing, Installing, and Improving a New Packaging System: Teamwork Between a Business and Its Customers

THIS CHAPTER describes how Mitsubishi's semiconductor equipment department works with its customers to develop, install, and improve a new generation of equipment; a new packaging system is used as an example. The packaging system integrates one die bonder, one mold machine, and a variable number of wire bonders depending on the package type.[1] The die bonder attaches the chips to a frame, the wire bonder attaches gold wires from the chip to the frame, and the mold machine creates a plastic package around each chip and its wires. This new generation of packaging systems also integrates several new elements of technology, some of which are described in Chapter 23 (see Figure 23-2). These technologies include a new bonding head, vision system, frame-feeding system, frame buffers between each equipment type, a quality control station, and software and hardware that integrate the three types of equipment into one packaging system.

As with the individual elements of technology, incremental improvement is an important strategy with the packaging system. The individual elements are installed in the packaging system as they are incrementally improved; moreover, the development and installation of the packaging system is not seen as a one-shot deal. Instead, Mitsubishi's semiconductor business looks at the project from the perspective of continual improvement; design modifications made in the initial installations are applied to the subsequent installations. The first installations were expected to require up to six months of modifications before they reached an acceptable level of yield (99 percent) and downtime (once per 2,000 chips). These design modifications were applied to the subsequently installed packaging systems, which were expected to reach the acceptable levels of yield and downtime much more quickly than the initial installations. This strategy obviously depends on close teamwork between the SED, the Mitsubishi semiconductor business, and the headquarters of the semiconductor business. As I'll describe in Chapter 26, other Japanese firms use a similar strategy.

Design, Manufacture, and Debug

Defining the Specifications

The packaging system was designed and manufactured using the methodology described in Chapters 17, 18, and 19. As discussed in Chapter 17, design engineers from the SED work directly with the equipment users to define the equipment specifications. In the case of the packaging system, engineers from the die bonder, wire bonder, and mold equipment groups worked with representatives from Mitsubishi's semiconductor factories and the IC assembly engineering department. Agreements were reached concerning these specifications in the assembly equipment communication meetings.

At the communication meetings, representatives of the organizations also agreed on a technology plan for the packaging system. This plan is similar to the plan described earlier for the wire bonder equipment except that it is for the entire packaging system (see Figure 23-1). This plan described the evolution of the packaging system in terms of the individual elements of technology. For example, the subsequent packaging model was expected to include one of the new bonding heads and vision systems described in Chapter 24.

These plans supported the incremental improvement of the packaging system. As shown in Figure 23-1, the elements of technology are usually tested by themselves, then within a stand-alone equipment type, and finally in the packaging system. Since the SED, its design documentation, the assembly equipment communication meetings, and the individual equipment communication meetings emphasize "modular design," it is very easy to incorporate these new elements of technology into the packaging system. Typically, a new element of technology can be introduced to the existing equipment simply by consulting the appropriate part of the equipment's documentation.

Translating the Specifications into an Equipment Design

The specifications agreed on in the assembly equipment communication meetings were translated into a design for the packaging system using the steps outlined in Chapter 18. The design documents were developed as the design proceeded and were stored in the public files for easy access by the members of the wire bonder, die bonder, and mold equipment groups. Macro-level equipment schedules were made by each equipment group's senior members, more detailed equipment schedules were made by more junior members of each group, and individual schedules were made by each engineer using standardized status forms. These schedules were also posted on the status boards so that everyone could follow the progress of the packaging system.

A good example of these standardized status forms can be found in the schedules used to develop the software and hardware that integrates the three equipment types in the packaging system. The wire bonder equipment group was responsible for developing the quality control station that contained the integrating software and hardware. It wrote detailed equipment schedules for the development of inter-equipment software modules such as production management, lot end processing, the communication between the high-level computer and the wire bonder's control system, and the communication between the various equipment types. These detailed project schedules and each engineer's individual schedules were updated each week and any problems that might influence these schedules were also included.

Problems with these software modules were discussed in various meetings. The weekly equipment meetings were used to determine which problems required more personnel to prevent a schedule slippage. Engineers were asked to modify their weekly schedules and overtime hours to solve the most important problems. Due to the product-oriented organization, design engineers reported to only one equipment group and it was easy for the equipment group leader to assign engineers according to that equipment's list of priorities.

The SED's formal communication networks also made it easy for each equipment group to assign priorities. Each assembly equipment's priorities are determined in the individual equipment and assembly equipment communication meetings. Representatives from each semiconductor factory and the IC assembly engineering department decide which shipments of equipment and which equipment problems are the most important. This helps the leader of each assembly equipment group to set priorities for each equipment order and equipment problem and to assign engineers accordingly.

The manufacturing section and the SED's suppliers became involved with the design of the packaging system very early in the design process. As discussed in Chapter 19, key subassemblies were tested and modified many times during the early phase of the packaging system design. The manufacturing personnel performed some of these experiments under the direction of the design engineers from the various equipment groups. A detailed cost analysis was performed on these key subassemblies as they were designed.

Installation and Incremental Improvement

In some ways, the delivery and installation of the first two packaging systems was a big event for Mitsubishi's semiconductor business. Several attendees of the assembly equipment communication meetings and several section managers from Mitsubishi's semiconductor factories spent a few days at the SED observing the packaging system's final tests. In other ways, however, the installation of the system was not so important, since the Mitsubishi semiconductor business was more concerned with the continual improvement of the system than with the system's actual performance immediately after installation. After installing the system, the SED and factory engineers involved with the project began working toward the system's three levels of yield and downtime goals. The responsibility for these goals was shared by the department, the headquarters of the semiconductor business, and the location of the system's first installation.

These organizations worked toward these goals via a philosophy of continual improvement: Identify the major problems and solve them one by one. Since most of the mechanical problems had been solved before the equipment was installed, the remaining problems primarily involved bugs in the equipment's software. Let's look at three ways that the SED worked with the semiconductor factory engineers to improve the packaging system's yield and downtime through eliminating these software bugs.

First, the engineers focused on the equipment interface problems. Engineers from each equipment group worked together to eliminate any bugs associated with interfaces between the three equipment types. According to Kiyonori Yoshitomi, who was involved with this part of the project, they used the system's external and internal specifications and HCP charts to solve these problems.[2] Because of the standard way in which these documents were prepared and their storage in the public files, they are very easy to use.

Second, these engineers focused on the software bugs associated with the individual equipment types in the packaging system. For example, Masayuki Hiroki was primarily responsible for the software that controlled the mechanical aspects of the wire bonder. He identified and eliminated the software bugs associated with mechanical problems such as frame movement or bonding action.[3]

Third, the engineers focused on the software bugs associated with individual elements of technology in an equipment type. For example, Naomi Yokoyama was primarily responsible for the wire bonder's vision system. Once the software bugs associated with the mechanical action had been eliminated, the vision system became the main cause of interruptions in the wire bonder's operation. Yokoyama would wait for the vision system to have such a problem, analyze the source of the problem, and revise the software so that a particular interruption would not occur again. Since most of the software bugs had been eliminated by this time, however, he often had to wait hours for a problem to occur. As he put it, "If you're unlucky, none of the problems occur while you're there."[4]

Teamwork Between Factories: Applying These Improvements to Subsequently Installed Packaging Systems

As these improvements (both software and frame-feeding system improvements) were incrementally applied to the initially installed packaging systems, they were also incorporated into the design of the subsequent packaging systems. Several additional systems were shipped before I left the SED and eventually this type of packaging system was expected to be the primary assembly method used in Mitsubishi's semiconductor factories. The subsequent systems would undoubtedly benefit from the lessons learned in the installation and improvement of the two initial packaging systems.

The first two packaging systems were probably installed earlier than a typical equipment installation in the United States due to the close relationship between the SED and Mitsubishi's semiconductor business. As stated earlier, these organizations did not view these first two installations as independent of the subsequent installations. They were concerned with moving new equipment technology into Mitsubishi factories as quickly as possible. Therefore, they wanted the equipment installed as soon as possible to begin working on all of the packaging system's production problems. The factory in which these initial systems were installed accepted these systems as part of its role in the Mitsubishi semiconductor business, although it would have obviously preferred to wait until the bugs were worked out of the equipment. It's always easier to be the second factory to use new equipment.

The subsequent packaging systems were installed more quickly and probably reached their yield and downtime goals more quickly than if the first system had been delivered after all of the problems had been completely solved in the SED's factory. Until semiconductor manufacturing equipment — or any product for that matter — is used in its working environment, it is impossible to forecast all of its potential problems. By installing the first system as soon as possible, the SED and the semiconductor factories were able to solve the problems and apply their solutions to the subsequent installations faster than if the first system had been delivered after all of its problems had been completely solved.

The key, however, to the early installation of this equipment is the close relationship between the SED and the Mitsubishi semiconductor business. Shared responsibilities and formal communication networks are important aspects of this relationship. Parts Four and Five described many examples of this close relationship and its benefits. Because of it, equipment can be standardized and common elements of technology can be used in different types of equipment. Standard material handling and control technology, both crucial to the implementation of CIM, are developed through this close cooperation. Since Mitsubishi develops a wide range of semiconductor equipment, it can also benefit from the use of common technology in its different equipment types.

In a similar manner, the Mitsubishi semiconductor business benefits by having a new semiconductor equipment model installed before it would normally be ready for delivery to outside firms. Early delivery enables the Mitsubishi semiconductor factories to receive new equipment before other semiconductor factories — those that do not have such a relationship with a semiconductor equipment business — get it. Mitsubishi's semiconductor factories can begin working out all of the equipment's bugs in a production environment before some of its competitors, particularly U.S. competitors, have received the equipment.

Japan's Horizontally and Vertically Integrated Manufacturing Firms: Teamwork Between Businesses

THE ACTIVITIES described in the previous chapters of this part are not unique to Mitsubishi's semiconductor equipment department and semiconductor business. The close relationship that exists between the SED and the semiconductor business is actually typical to most Japanese firms. Prestowitz, Ferguson, Schonberger, and the MIT Commission on Industrial Productivity all note the close relationships that Japanese firms develop with their internal and external customers and suppliers. Although none of these authors has described the linkages between these businesses in the detail presented here, there is a great deal of evidence that the typical linkages between Japanese businesses are very similar to what I have described.

I described the linkages between SED and the Mitsubishi semiconductor business in terms of shared responsibilities and formal communication networks, concepts that are very similar to William Ouchi's concepts of collective responsibility and collective decision making. Ouchi maintains that the concepts of collective responsibilities and collective decision making characterize the way that organizational groups within a

Japanese firm (including customer and supplier relationships) allocate responsibility and make decisions.[1] Therefore, close relationships between two businesses, departments, or sections in a Japanese firm are likely to be similar to the relationship I have described. Let's see how common horizontally and vertically integrated firms are in Japan and discuss the advantages and disadvantages to close relationships between a business and its customers and suppliers, as well as which products most need these close relationships.

Integration in Electronics Firms

Several observers have noted that Japanese electronics firms are horizontally and vertically integrated. They are vertically integrated in that many of their divisions supply equipment or electrical components to the divisions that produce final electrical products. They are horizontally integrated in that they produce a wide variety of electrical products that are related by common technologies. For example, six companies — Hitachi, Fujitsu, NEC, Toshiba, Mitsubishi, and Matsushita — control most of Japan's semiconductor, computer, telecommunications equipment, and consumer electronics production. They are vertically integrated in that most of them produce their own chips and some of their own semiconductor equipment or have a close relationship, which may include an equity interest, with a semiconductor equipment firm. Mitsubishi produces most types of semiconductor equipment. Toshiba and Hitachi produce photolithographic equipment. Toshiba has a close relationship with Tokuda Seisakusho, which produces etching equipment. Hitachi has a similar relationship with Showa Denko, which also produces etching equipment. Hitachi also produces ion implantation equipment. NEC has a close relationship with firms that produce chemical vapor deposition equipment (Anelva) and test equipment (Ando Electric). Fujitsu produces etching equipment and has a close relationship with two producers of test equipment (Advantest and Takeda Riken).[2]

These six dominant firms are also horizontally integrated. Each of them also produces various types of consumer electronics equipment. Mitsubishi produces televisions, VCRs, stereo equipment, large-screen projection systems, and compact disc players. NEC produces televisions, VCRs, video disc players, portable video cameras, tape recorders, hi-fi audio systems, compact disc players and videotext terminals. Toshiba produces televisions, picture tubes, liquid crystal displays, VCRs, and audio and video systems. All of these firms are developing high-density televisions (HDTVs).

Each of the six firms produces telecommunications equipment and most of them either produce optical fiber or have a close relationship with a producer of optical cable. For example, Mitsubishi produces radar systems, mobile telephones, and facsimile machines; it has a close relationship with Dainichi Cable. Toshiba produces facsimiles and has a close relationship with Fujikawa Cable. NEC produces digital switching systems, facsimiles, mobile communication systems, and integrated switching digital networks. It is the largest supplier of private branch exchange systems in Japan and the fourth largest supplier in the United States. It controls 30 percent of the satellite dish market. It has a close relationship with Sumitomo's optical fiber division.[3]

Each of the six firms also produces computer systems and peripherals. For example, Fujitsu is the largest computer maker in Japan and the third largest in the world if its linkup with Amdahl is included. NEC is the second largest computer maker in Japan and the fourth largest in the world. It supplies 56 percent of the personal computers in Japan and is the fourth largest supplier of PCs in the United States.

Other electronics companies in Japan are also horizontally integrated as compared to most U.S. electronics firms. Canon produces photographic, telecommunications, and semiconductor equipment, as well as office products. Sony produces consumer electronics, semiconductors, semiconductor wafers, industrial electronics; through its purchase of Columbia Pictures, it hopes to display the benefits of its HDTVs to the American public.

Most of these other Japanese electronics companies are also vertically integrated. Canon, Fuji-Xerox, and Epson are just a few companies that have close relationships with their parts suppliers.[4] More important, no one has identified a successful Japanese electronics firm that does *not* have close relations with its suppliers.

Other Examples of Integrated Manufacturing Firms

Several other Japanese industries are also dominated by firms that are vertically and horizontally integrated. Michael Cusumano's study of the Japanese automobile industry found that the Nissan and Toyota groups both have developed extensive networks of parts, assemblies, and equipment suppliers. He notes that there is close cooperation between each of these companies and their respective parts and equipment suppliers.[5]

The MIT Commission on Industrial Productivity also concluded that Japanese manufacturing firms have developed close relationships with their customers and suppliers. In particular, it found that Japan's machine tool industry was less fragmented than the U.S. machine tool industry. The Japanese machine tool firms are specialized and most have close relationships, often in the form of equity ownership, with automobile or other large manufacturing firms. The commission also found a close cooperation between these machine tool producers and Japan's largest producers of NC controls, FANUC (a subsidiary of Fujitsu). The commission argued that the close linkages between the Japanese machine tool industry and FANUC have enabled Japanese manufacturing firms to use equipment with standard controls, which has facilitated the integration of manufacturing equipment into computer-integrated manufacturing systems.[6]

This is similar to what I found in Mitsubishi's semiconductor business. The close linkages between the Mitsubishi semiconductor equipment department and its semiconductor business are facilitating the implementation of computer-integrated manufacturing in its semiconductor factories. A key aspect of these efforts is the joint development of a standard equipment controller and semiconductor factory production system.

Michael Porter makes a similar argument about a specific type of machine tool, industrial robots. He observes that Japan's robotic industry depends heavily on tight linkages between producers and users of robots and between these robotic producers and the producers of industrial controls. According to Porter, most major manufacturing firms in Japan produce robots to meet their own special-purpose needs. Japan's 130 largest robotic makers are producers of electrical equipment, machinery, transportation equipment, or steel. Porter also argues that Japanese robotics makers benefit from their involvement with electronics and computers. They have an advantage over machinery companies in incorporating sophisticated electronics and controls in their machinery.[7]

The Role of MITI

The Ministry of International Trade and Industry (MITI) also supports horizontal and vertical linkages between Japanese businesses by promoting cooperative research between firms. By developing research consortiums that include multiple firms in an industry, MITI helps Japanese industry share the cost of new technology between firms and promotes cooperation between the developers and users of new technology.

One of the key reasons for MITI's success in promoting these types of cooperation, however, is that Japanese firms are already accustomed to a great deal of *internal* cooperation. Japanese firms are used to close cooperation between a department and its internal customers, suppliers, and laboratories. Japanese firms also use sophisticated corporate R&D selection processes to develop, integrate, and modify the plans for research projects. Without this type of internal cooperation, it is doubtful that MITI's projects would have much success.

For example, an important aspect of MITI's VLSI project depended on close relations between a producer and a user of photolithographic equipment. MITI chose Nikon, already a successful producer of precision optical equipment, cameras and lenses, to cooperate with Toshiba, one of Japan's largest electronics companies, on the development of photolithographic equipment. The Office of Technology Assessment, a research arm of the U.S. Congress, concluded that close relations between Nikon and the users of this equipment were the main reason why Nikon was able to pass GCA in the late 1970s (GCA was formerly the world leader in photolithographic equipment).[8] However, as the earlier description of Mitsubishi's semiconductor business shows, close relationships between semiconductor equipment producers and users is a natural arrangement in a Japanese firm. MITI merely promoted a relationship between a supplier and a user of photolithographic equipment that did not previously exist.

A second example of MITI promoting close relations between producers and users of a new technology is in the area of machine tool controls. According to the MIT Commission on Industrial Productivity, MITI supported FANUC's development of standard machine tool controls that were compatible with each other — unlike development in the United States, where incompatible controls were developed by a fragmented machine tool industry.[9] However, as the earlier description of Mitsubishi's semiconductor business also shows, close relations between the producers and users of equipment controls (in this case between the developers of standard equipment controls and factory system controls) is a very natural arrangement. As with the VLSI project, MITI promoted a type of relationship between the producers and users of a new technology that Japanese firms were already very accustomed to.

Examples of MITI's promoting horizontal linkages between firms also exist. For example, MITI supports research consortiums in the areas of superconductivity[10] and micromachines (motors, sensors, and other devices that are built on semiconductor chips using various semiconductor technologies).[11] Both of these technologies are very expensive to develop, but MITI's support of the research consortium and the cooperation between multiple firms make them more affordable to individual Japanese firms.

Horizontal integration and a custom of sharing technologies across different businesses further lower the cost to individual firms. We saw how Mitsubishi uses a sophisticated corporate R&D selection process to develop, integrate, and modify the plans for individual research projects. We also saw how Mitsubishi developed close relationships between laboratories and divisions. MITI's research consortiums merely add an additional level of cooperation to industries that already have firms that support cooperation between different businesses and between laboratories and divisions.

MITI, therefore, is not the prime cause of horizontal and vertical integration in Japanese firms; Japan's corporate culture is the primary reason why Japanese firms have succeeded at developing and benefiting from these linkages. MITI, however, does help Japanese firms develop new linkages that firms feel are beneficial.

Benefits of Horizontal and Vertical Integration

Many Western observers have focused on the negative aspects of Japan's horizontally and vertically integrated firms, particularly their unwillingness to buy foreign products. These integrated firms are more likely to buy a product that is produced internally or by a firm in the same economic group than to buy from a U.S. firm or even another Japanese firm, because individual departments value long-term relationships with customers and suppliers.

To many Americans, not only is this protectionism, but it also appears to be inefficient. According to free-market theories, firms should purchase raw materials and equipment from the cheapest source. Each purchase should involve the evaluation of several suppliers, with individual purchases basically independent of each other. This philosophy, however, ignores the benefits from having long-term relationships between "related and supporting industries."[12] These relationships are not only between a business and its customers and suppliers — they are also between businesses that produce related products. Some of these benefits were discussed previously, using examples from the relationship between Mitsubishi's semiconductor business and its semiconductor equipment department. It is useful to review these benefits and those reported by other observers to identify which products require tight linkages between a business and its customers and suppliers.

Michael Porter provides the most comprehensive summary of the benefits from tight linkages between related and supporting industries.[13] Three of the benefits are ongoing coordination in the value chain (i.e., production process), innovation and upgrading, and shared activities in the value chain.

Coordination in the Value Chain

With respect to the first benefit, Porter notes that linkages between the value chains of firms and their suppliers are important to competitive advantage.[14] As discussed earlier, new manufacturing strategies such as JIT production and concurrent engineering depend strongly on tight linkages between firms and their suppliers.

Several Japanese authors have also argued that Japan's vertically integrated economic groups and *keiretsu* (enterprise unions) enabled Japanese firms to survive the increasing value of the yen in the mid-1980s. When the value of the yen increased by more than 50 percent in late 1985 and early 1986, many Japanese firms needed to reduce their costs by a similar amount in order to stay in business. According to several Japanese authors, the existence of tight relations between successive links in the customer-supplier chain helped bring about these cost reductions. Producers of finished products requested that each of their suppliers reduce their costs by a certain percentage. These first-tier suppliers then asked the second-tier suppliers to reduce their costs by a similar amount, and so on. Each supplier reduced its costs via productivity improvements that enabled the finished producers, and thus all of their suppliers, to maintain their market share.[15]

Innovation and Upgrading

Innovation and upgrading is a second benefit of tight linkages with suppliers and customers. Porter argues that suppliers and customers can help firms identify opportunities to apply new technologies. Firms can influence the supplier's technical efforts and customers can provide test sites for development efforts. The exchange of technical information and joint problem solving also reduce the cost of R&D to each firm.[16] In a similar vein, Eric von Hippel comments that lower risks are associated with the financing and implementing of new technology by users (as opposed to suppliers) because users control the market for the technology.[17]

Many of these benefits were found in the relationship between Mitsubishi's semiconductor equipment department and its semiconductor business. The SED helps the semiconductor business identify new opportunities in wafer fabrication. The semiconductor business influences the department's technical efforts; it helps write the plans for developing the elements of new equipment technology. The semiconductor business is also a test bed for the SED's products: Packaging equipment is installed in a semiconductor factory before it is working perfectly; the bugs are worked out through incremental improvements. Joint problem solving was used in developing standard equipment controllers that can be easily integrated with the computer-integrated manufacturing systems used in the semiconductor factories.

Other studies have found similar results. For example, MITI's Small and Medium Size Enterprise Agency conducted a survey of Japanese manufacturing subcontractors. Some 45 percent of respondents said they received technical assistance from a parent company, 37 percent received information, 28 percent were loaned or leased equipment, 24 percent received training for their employees, and 14 percent received financial assistance.[18] Each of these subcontractors probably had a somewhat stable market for their product, which is necessary to implement new production technology. In a study of U.S. metal working firms, about half the firms that had not bought numerically controlled or computer-numerically controlled machine tools cited the lack of stable demand for their product as the reason.[19]

Shared Activities in the Value Chain

A third benefit to tight linkages is that related industries can share or coordinate activities in the value chain. These activities can be shared in manufacturing, distribution, marketing, or technology development.

Several examples of shared activities were described previously; some are within the semiconductor equipment department and others are between different departments within Mitsubishi. Within the SED, standard software, hardware, mechanical technologies, and vision technology are defined and used to reduce equipment costs and improve equipment quality. As for sharing technology between different departments, Mitsubishi's semiconductor business standardizes the equipment needs of multiple factories to reduce the cost of new equipment and of integrating this equipment into computer-integrated manufacturing systems. Mitsubishi also integrates project proposals in its corporate R&D selection process to spread the cost of developing new technology across different electrical industries.

The Triple Disadvantage of U.S. Firms

Japan's horizontally and vertically integrated manufacturing firms provide U.S. firms with a triple disadvantage in terms of U.S.-Japanese trade. First, these tight linkages provide Japanese firms with a competitive advantage. Second, they discourage U.S. imports. Many Japanese firms will buy products only from suppliers with whom they have developed long-term relationships.

Third, these tight relationships prevent U.S. firms from obtaining many types of equipment and components as early as Japanese companies can get them. As described previously, Japanese firms like to transfer new technology between firms before all of the bugs have been worked out. By transferring the technology early, Japanese firms can begin working out the bugs much more quickly than if they waited for the supplier to solve the problems on its own. However, Japanese firms transfer technology in this manner only when they have a close relationship between the supplier and user of the technology. Since U.S. firms are generally not part of these close relationships, they obtain the new technology (frequently in the form of new equipment or new components), after the Japanese firms get it.

Semiconductor, computer, and automobile firms in the United States are already experiencing such delays. Semiconductor firms have reported a lag in obtaining new semiconductor equipment.[20] Cray and other U.S. computer makers have reported delays and higher costs in receiving semiconductor chips.[21] General Motors claims that production equipment it purchases from Japanese equipment makers has already been in use at Toyota for two years.[22] Other U.S. industries are probably also experiencing similar delays and are similarly disadvantaged with regard to these linkages.

It's clear that there are a number of advantages to close relationships between firms in related and supporting industries. However, some industries require more horizontal and vertical integration in their firms than other industries, and some products within these industries require more integration than other products. First, let's look at those products and industries that require the most integration, then Part Seven will use these observations to forecast the industries in which Japan will be the most successful in the future.

Products that Require Horizontal and Vertical Linkages

As discussed in Part Two, teamwork has increased in importance over the last 70 years due to products' having a greater number of parts, more product diversity, more complex components, a greater use of automation, a greater number of different technologies, and shorter product lives. Japanese firms have been evolving from low-technology to high-technology industries during the last 20 years. When this technological evolution is taken into account, Japanese firms are doing particularly well in the industries that require the most teamwork. As noted in Part One, Japanese firms are doing better in the discrete parts manufacturing industries and in those industries that have a large number of manufacturing steps, areas that require more teamwork.

Japanese firms also do better in industries and products that most benefit from horizontal and vertical integration because this type of integration relies on teamwork. The reasons for the importance of horizontal and vertical integration to specific industries and products are related to some of the factors described in Part Two. Two of these factors are particularly important: greater product diversity and greater component complexity.

Specialized Products

Specialized products, which are becoming more prevalent due to increased product diversity, require more vertical integration than do nonspecialized products. Specialized products, and industries that use them, require close feedback between the producers and users to design the products and develop new technology for the products. Examples include key subassemblies for a complex product, advanced materials such as ceramics or composites, or manufacturing equipment. Each of these products is typically customized for a specific firm.

Manufacturing equipment is particularly important due to the increased complexities of today's manufacturing processes. Tolerances are tighter, setup times are shorter, and equipment must be integrated into computer-integrated manufacturing systems. The result is that more and more manufacturing firms are depending on their equipment suppliers for customized solutions to their manufacturing problems, and a manufacturing firm's competitiveness is becoming more dependent on the capability of its manufacturing processes.

Complex Components

Complex components also require horizontal linkages between related products and vertical linkages between producers and users of the components. Horizontal linkages are required due to the high cost of developing many complex components. Many types of components can be developed only by firms that have a wide variety of potential applications for these components. For example, new semiconductor devices such as gallium arsenide field-effect transistors, micromachines, or three-dimensional chips are extremely expensive to develop and thus require large potential markets to make investment in these technologies worthwhile. The process technology required for a new memory chip costs about $100 million to develop. Only firms that can apply the technology to a wide variety of chips can afford this level of investment.[23]

Vertical linkages are also needed between the producers and users of these complex components because many of these components are becoming more specialized. For example, plastic and metal parts are increasingly dependent on special materials and manufacturing processes. Semiconductor devices, particularly application-specific integrated circuits (ASICs), require teamwork between the producers and users of these components to develop the appropriate circuit design. Therefore, close relationships are needed between the producers and users of these components to identify the most appropriate materials and processes.

However, the best example of complex components that are also specialized is electrical products. Today's electrical products are becoming increasingly specialized. Further, due to the increasing "integration" of these products, most electrical products are becoming components within complex electrical systems. Computer-integrated manufacturing, office automation, home automation, and telecommunication systems all require the integration of various products or components. Even consumer electronics, particularly "multimedia" televisions, computers, and VCR-based photo imaging equipment, will eventually be part of these home automation systems.

Actually, most products require a firm to have tight linkages with its customers and suppliers. The notable exceptions are high-volume, standardized products that do not depend on specialized components. These products have a wide variety of users either in the commercial or industrial markets. Although the firms producing these products must keep their sights on the customer, they may not need equity relationships with their customers or a strong customer role in product design. These firms will continue to depend on broad marketing analyses to identify customer requirements and design products that meet those requirements.

Some readers may be thinking about Japan's successful producers of high-volume standardized products. For example, the automobile and consumer electronics industries are both high-volume industries in which Japanese firms have been highly successful. To become successful, however, these industries have depended on close relationships with their suppliers of specialized equipment and components. Japan's automobile industry depends on special-purpose equipment and parts from its network of suppliers. Its consumer electronics industries depend on special-purpose components from its suppliers, such as charge-coupled devices, liquid crystal displays, and more general-purpose devices such as television picture tubes, optical elements, and memory chips.[24] In the mid-1980s, 12 percent of Japan's production of memory chips were used in its VCRs.[25]

The best example of a consumer product that requires specialized components is high-density television (HDTV). The development of HDTV will require extensive cooperation between component suppliers and the producers of the televisions. It probably will also provide innumerable spinoffs in related businesses. Many observers believe that the large market for HDTVs will enable component suppliers to finance the development of several specialized components that will have many other applications. Jeffrey Hart and Laura Tyson note that HDTV will require the development of several specialized components, such as liquid crystal displays and light-emitting diodes for flat panel displays and chips that perform digital signal and image processing. Further, HDTVs will provide a market and stimulus for innovation in the development of interactive video, interactive 3-D color, optical fibers, and networks. Hart and Tyson believe that these technologies will also be applicable to computers, particularly engineering workstations and telecommunications.[26]

Less Linkage-dependent Products

Some industries do not require as much horizontal and vertical integration, and U.S. firms tend to do well in those industries. These types of industries produce high-volume standardized products that do not depend on specialized components. For example, as noted in Part One, U.S. firms continue to be successful in chemicals, plastics, and pharmaceuticals; these industries depend less on specialized components than do the discrete parts industries. The materials used by these industries are primarily petroleum or other raw materials that are highly standardized.

Michael Porter provides additional evidence for the relative unimportance of vertical linkages between the final producers of chemicals, plastics, and pharmaceuticals and their suppliers of specialized machinery and materials. Porter classifies all products, services, raw materials, and foodstuffs into 16 major clusters; each cluster includes relevant machinery, specialty inputs (materials), and services. He finds, almost without exception, that for a country to be competitive in a cluster, the cluster must include competitive suppliers of machinery, materials, and services. However, chemicals, plastics, and pharmaceuticals are notable exceptions to this rule. In the major producers of these products — the United States, Japan, and the former West Germany — Porter did not find significant exports from any machinery, raw material, or service industry that also provided a specialty input to the chemical, plastics, or pharmaceutical industries.[27]

Contrast this with the automobile industry. According to Porter, the Japanese automobile industry is fed by 27 different machinery and parts industries in which Japan has a significant fraction of world exports. The automobile industry in the former West Germany is fed by 21 different machinery and parts industries in which that country has a significant fraction of world exports. In other words, the automobile industry requires close relationships between the automobile producers and their suppliers; the chemical, plastics, and pharmaceutical industries, on the other hand, do not require close linkages between the firms in these industries and their suppliers because they don't depend on specialty inputs.

The relationship, then, between Mitsubishi's semiconductor equipment department and its semiconductor business is not unique. Many Japanese industries have close relationships between the firms in these industries and their customers, suppliers, and related industries. Although this is particularly true in the electronics industries, these types of relationships have also been found in the automobile, machinery, and robotics industries. Further, these relationships appear to provide Japanese firms with a competitive advantage in the discrete parts industries while they appear to be less important in the nondiscrete parts industries.

CHAPTER TWENTY-SEVEN

Long-term Relationships and Long-term Thinking

EVERY DISCUSSION about Japanese manufacturing mentions the apparent long-term orientation of Japanese firms, particularly in comparison with U.S. firms. Japanese firms invest more time and resources in new technology than do U.S. firms; Japanese firms, for example, are investing more funds in developing superconductivity and HDTV.[1] Japanese firms are also more interested than their U.S. counterparts in employee training programs. And Japanese firms develop long-term relationships with their customers and suppliers.

The conventional reason for this difference in outlook is that U.S. firms are more constrained by the stock market than Japanese firms. American managers can't look at anything but the bottom line and the next quarter's profits, because they need to satisfy the stockholders. Japanese firms aren't so constrained by the stock market; they have access to lower cost funds and have cross-ownership of shares with banks and other companies. Therefore, when U.S. managers use discounted cash flow analyses to evaluate investments in technology, new products or training programs, they use a small number for the life of the project and a higher discount rate than Japanese firms.

Because Japanese companies appear to have access to lower cost capital and are less beholden to the stock market, it makes it easier for them to invest in new technology, new products, and training programs. However, since the long-term orientation of Japanese firms is reflected in such a wide variety of issues, it is hard to believe that the differences in outlook are due only to differences in the Japanese and U.S. stock markets. There are two problems with the conventional argument.

First, do Americans believe that U.S. firms could compete with Japanese firms in such markets as superconductivity or HDTV even if the investment funds were available? Japanese firms have been beating U.S. firms in several high-technology industries for many years. As noted here and in many other studies, much of Japan's success can be traced to things that are completely independent of the stock market: the ability to implement such strategies as JIT production, fast-cycle-time product development, and the fast commercialization of new technology. The reluctance of U.S. firms to invest in these new technologies may be based more on an intelligent understanding of the situation than on simple shortsightedness.

Second, and more important, does the top management of Japanese firms have enough power to dictate the activities and decisions made by middle and lower management? If differences in the way stocks are traded in the two countries are the reason for the long-term orientation of Japanese firms, Japan's top management would have to have a great deal of influence on how decisions are made deep in the organization. Since most of the decisions concerning new technology, new products, or new manufacturing systems are made by middle and lower management in U.S. *and* Japanese organizations, Japan's middle and lower management must be greatly influenced by the top management if differences in the U.S. and Japanese stock markets are the actual reason for the long-term orientation of Japanese firms.

However, many observers of Japanese management feel that bottom-up management is more widely practiced in Japanese firms than in U.S. firms. Whether Japanese firms actually use a completely bottom-up approach to these issues is not the question (I don't believe they do). The issue is whether Japanese firms use a more bottom-up approach than U.S. firms, and it is fairly clear that they do in that many studies have found that Japanese firms are more decentralized than U.S. firms.[2]

A Long-term Orientation Based on Japan's Corporate Culture

It appears that Japanese firms do many things that reflect a long-term orientation as an indirect result of their corporate culture. Earlier chapters described numerous examples of this connection between Japan's corporate culture and Mitsubishi's long-term orientation. Most of these examples are related to close intergroup relationships and incremental improvement of products and processes. Relationships between groups are defined by two subelements of Japan's corporate culture: shared responsibilities and formal communication networks. Incremental improvements in products and processes are primarily implemented via another subelement: employee developed procedures and plans.

Close intergroup relationships and incremental improvements in products and processes are also related to each other. Incremental improvements require close relationships between the various groups involved with the products and processes to provide fast feedback concerning these improvements. Close relationships between groups also encourage these incremental improvements and keep the groups moving in the same direction. Let's review several examples presented earlier to see how close intergroup relationships and incremental improvements in products and processes are related and how these factors reflect a long-term orientation within Mitsubishi.

Relationships Between Businesses

Japanese firms develop long-term relationships between different businesses, departments, and sections within a firm, as well as between firms, to share risk and to focus on long-term growth. These long-term relationships are established because Japanese firms believe in these types of relationships, not because they have done an economic analysis that has shown them to be beneficial. They believe in these long-term relationships because Japan has emphasized these types of relationships throughout its history.[3]

Relationships between banks and firms within economic groups are one example of these long-term relationships. These banks and firms own stock in each other, and few shares are traded on the stock market. As discussed above, many observers believe that these relationships are the main reason for the long-term orientation of Japanese firms. These, however, are but one example of many such relationships that exist at all levels in a Japanese manufacturing firm.

Several examples have been provided. First, the Mitsubishi semiconductor business and the semiconductor equipment department have developed a long-term relationship in which the SED supplies the semiconductor business with most of its equipment. The semiconductor business provides the SED with a stable market for its equipment in return for influence over the department's design and technology development decisions. Long-term plans are agreed on in which the semiconductor equipment is incrementally improved. The semiconductor business provides a test bed for the semiconductor equipment, employees are exchanged between the two organizations, and joint problem solving occurs.

Second, the SED and the semiconductor business have developed long-term relationships with several of Mitsubishi's laboratories, particularly the LSI laboratory. The LSI laboratory provides the SED with a stable market for its new equipment in return for influence over the department's design and technology development decisions. The LSI laboratory can make this type of long-term agreement with the SED because the semiconductor business provides the laboratory with a stable research budget in return for influence over the laboratory's technical decisions. Long-term equipment plans are agreed on in which the equipment is incrementally improved. Employees are exchanged between the three organizations and joint problem solving takes place.

Third, even Mitsubishi's corporate R&D selection process reflects long-term relationships between different businesses within the company. Individual project proposals are integrated with other proposals, modified based on other proposals, and reintegrated with these proposals in this selection process. Further, many of these project proposals are based on product development requests that are jointly developed between different businesses. Through these product development requests and by integrating the individual research proposals, Mitsubishi is able to identify projects, particularly large projects, that have a wide variety of potential applications. Since the cost of these large projects is shared across several businesses, there is less risk in funding these large projects.

Relationships Between Sections Within a Department

Similar examples of long-term relationships in which there is incremental improvement of products and processes can be found at the departmental level in a firm. Long-term relationships are established between different sections within a department to incrementally improve a product or a process. I have described several examples of how these relationships support incremental improvements of employee developed procedures and plans in the semiconductor equipment department. These long-term relationships were supported by shared responsibilities and formal communication networks.

Part Five described how the SED uses shared responsibilities and formal communication networks to incrementally improve its design systems. Cycle time is reduced and quality is improved through incremental improvements in the department's design methodology. The software design methodology, which included detailed procedures and documentation, was particularly emphasized. Improvement plans are developed in which incremental improvements to these procedures and documents are related to long-term investments in new software design tools.

Incremental improvements in procedures are even stressed in the use of CAD. Instead of just buying a large number of CAD stations and having each engineer learn to use these machines on their own, the SED emphasizes the development and improvement of a procedure for using the CAD stations before it will make a substantial investment in CAD equipment (see Part Five). According to the reasons conventionally given for Japan's long-term orientation, the department should have purchased this equipment as soon as possible, since the department has access to low-cost capital. However, the department believes that in terms of long-term performance, it is better first to incrementally improve its CAD procedures.

The SED also looks at the performance of its equipment in a similar manner. Instead of emphasizing large, risky improvements in performance, the department focuses on incremental improvements in each equipment's performance and in the technological elements of each equipment type. In conjunction with the semiconductor business and laboratories, long-term plans are developed that describe these incremental improvements.

Lower interest rates and higher debt-to-equity ratios have little to do with the SED's emphasis on incremental improvements in products and processes or on close relationships between different sections in the department. If interest rates and debt-to-equity ratios were the key factors, Japanese firms would emphasize purchases of CAD equipment, software tools, and large jumps in technological performance rather than incremental improvement in CAD procedures, software procedures, and product performance.

The examples presented here suggest that Japanese firms don't develop long-term relationships between different firms or between different businesses, departments, or sections within a firm because they have justified them economically. They develop these relationships because Japan's corporate culture compels them to behave this way.

Japanese firms don't emphasize incremental improvements in manufacturing systems, design systems, and products based on an in-depth economic analysis. They emphasize incremental improvements because this works best in their group-oriented corporate culture, in which relationships between groups and individuals are emphasized. In order for individuals and groups to have constant input, systems and products must be improved in small increments and detailed plans must be used to guide their actions. Japanese firms don't use economic analysis to justify other elements of their corporate culture. The product-oriented organization, visible management, multifunctional workers, open offices, and extensive training were never economically justified. Japanese firms do these things because their culture tells them to.

In a similar manner, U.S. firms don't use economic analysis to justify many elements of their corporate culture. Economic analysis is not used to justify the use of individual offices, personal secretaries, functional organizations, individual responsibility, private files, private information, or functional specialization. Firms in the United States do these things because Americans believe in them.

Some time ago I spoke with a group of American engineers who were developing ceramic-based fuel cells. Some of these engineers had been involved with fuel cell development for more than 20 years; in the last few years, they have begun to sell a few cells (this is definitely a long-term project). We were talking about the importance and difficulties of processing the ceramic material used to produce the fuel cells. I asked these engineers if they had considered developing a long-term relationship with a ceramics company to develop better manufacturing processes for these ceramic fuel cells. They replied that this type of relationship wasn't necessary since they read all the appropriate journal articles and know what's going on in the ceramics field. At another point in the conversation, these engineers criticized the shortsightedness of their parent company for not investing more in R&D. I simply agreed, unable to bring myself to point out their apparent contradiction in logic.

PART SEVEN

Japan's Weaknesses and Lessons for the United States

Japan does provide lessons for the United States, particularly with respect to teamwork. The United States needs to become much better at implementing just-in-time manufacturing and concurrent engineering, reducing product development time, working closely with suppliers and customers, and shortening the time it takes to commercialize new technology. Japan has been successful in these areas because its corporate culture supports the teamwork needed to achieve these goals. Aspects of this corporate culture can be applied to American companies.

Japan, however, is not perfect. Japan has many problems, including difficulties that relate directly to its corporate and social cultures. Before making recommendations about the application of elements of Japan's corporate culture to U.S. firms, it's important to examine these problems.

CHAPTER TWENTY-EIGHT

The Downside of Japanese Management and Culture

A NUMBER OF Western authors have commented on negative aspects of Japanese society and Japanese management. Karel van Wolferen, Jared Taylor, and other writers paint a picture of a society with seemingly few redeeming features: a society with a low standard of living, an inefficient service sector, abuses of individual rights, sexism, and racism.[1] After reading these books, one might wonder how Japan has become so successful. These books represent views that are typical of Americans who have spent a considerable time in Japan. Most Americans living in Japan whom I have met have similar views; they seem either unaware of Japan's manufacturing success or unable to connect their view of Japan with Japanese economic success. This perspective may seem surprising until you consider that most Americans living in Japan are missionaries, English teachers, or employees of Japanese subsidiaries in the United States. These people rarely get the opportunity to see any evidence of Japan's manufacturing prowess.

One might think that Americans employed by Japanese firms would have an opportunity to develop a great deal of respect for Japanese management and society. I've found that most of these employees, however, have much stronger feelings about Japan that prevent them from being impressed by Japanese companies. Some of these employees feel like Jay Edwards (the hypothetical character introduced in Part One, who was confused by Japanese offices and management practices); others see only the relatively crowded living conditions and high prices or are overwhelmed by the Japanese language, the lack of individual freedom, or, in many cases, the blatant racism.[2] I bring up these views for two reasons: First, many American visitors to Japan whom I have met seem to feel this way about Japan, and second, many of the negative aspects of Japan are related to its corporate social cultures — the very things I argue help Japanese manufacturing firms be more competitive.

A Low Standard of Living

Japan has seen its standard of living rise incredibly fast over the last 50 years. In fact, it has risen faster than that of any other country in the world's history, although it still trails the United States and many Western European countries. Even if Japan continues to increase its manufacturing productivity faster than the United States, which is likely, Japan's standard of living may take years to catch up with that in the United States because of Japan's high land prices and its lagging service and agricultural sectors.

With respect to high land prices, although much of this is due to the shortage of usable land, some of it is due to culture. Japanese people like to live close to each other in order to be associated with a group or groups. Riders of the Shinkansen (bullet train) in Japan see farmhouses tightly packed together, as opposed to families living adjacent to their own rice fields. This type of behavior also extends to the suburbs of Japanese cities. In Fukuoka, my Japanese friends would not even consider living just outside the city, though this land was less expensive and closer to where they worked. They wanted to live in the city to be closer to other people. The result is that there is a great deal of inexpensive and unused land outside of Fukuoka, some of which is adjacent to the ocean.

Although Japan's high food prices are primarily due to its lack of good agricultural land, they also stem from cultural influences. Van Wolferen argues that Japan will not significantly increase imports of food in the near future due to the way in which power is shared among various economic constituents. According to van Wolferen and others, Japan has little interest in such principles as free trade. It is more interested in maintaining the balance of power between its key economic constituents.[3]

This philosophy appears to be culturally based. It is similar to the long-term customer-supplier relationships among Japanese manufacturing firms. These firms do not change suppliers merely because another supplier (particularly one from another country) has lower prices or even higher quality than existing suppliers. Instead, responsibilities are shared between the firm and its suppliers and customers for reducing costs and improving quality. In the case of agricultural products, however, Japan does not have the capability, barring significant breakthroughs in agricultural science, to compete with imported food no matter how hard its farmers work together and with their customers.

Japan also has high prices for food and other manufacturing goods due to a dysfunctional distribution system. Products from the United States are much more expensive in Japan than in the United States, but the Japanese distribution system is so bad that even products manufactured in Japan cost more there than in the United States. Van Wolferen argues that Japan's distribution system doesn't work because of laws that protect small shops and because of long-term, basically monopolistic relationships between manufacturing companies, distributors, and retail outlets.[4] This protectionism is an aspect of power sharing between various economic interests. These monopolistic relationships represent the downside of long-term relationships and shared responsibilities between a firm and its suppliers and customers. Although these long-term relationships and shared responsibilities help make the final manufacturing firm in the value chain extremely competitive in the international marketplace, a significant lack of competition at key points in the value chain seems to prevent the service sector from improving its productivity.

Rights of Individuals

Another negative aspect of Japanese culture is its relative lack of individual rights. Japan expects its citizens to sacrifice more of their personal rights for the good of their companies and their political system than Westerners are expected to relinquish. For example, Japanese employees are expected to work a great deal of overtime, which can be thrust on them without warning or consent. Close relationships between a business and its suppliers, customers, and laboratories require a great deal of travel, making business trips frequent and often spontaneous. Family-related matters are expected not to interfere with these business trips.

Employees also have very little control over their type of work or place of employment after entering a firm. Not only is it rare for employees to request and receive transfers to other jobs or locations, it also rare for them to refuse such transfers. In addition, regular employees of large Japanese firms have many more rights and opportunities than employees of the small firms that supply parts and assemblies to the larger firms. Employees of these suppliers are considered second-class citizens to the employees of the parent firm. They have little opportunity for advancement because the top management of the supplier firm is largely drawn from the parent company. Employees of these suppliers typically make less money, and their jobs are less secure than jobs in the parent company. Their only crime may have been to graduate from a second-tier university — or they may have demonstrated a "lack of dedication" by changing companies.[5]

These are just a few examples of how life in Japan, particularly in a Japanese company, is not expected to be fair to the individual. It is also the downside of having your career depend on the group you are associated with. Although this exists in all cultures, it is especially pronounced in Japan, particularly with the problems with leaving a group. No matter how well a supplier performs, its employees will always be identified with that group. It is difficult to change groups; a person can only rise or fall with the group. Although this is typical of most Japanese companies, employees of second- and third-tier suppliers are particularly restricted in terms of advancement due to the way in which suppliers are economically trapped by their parent companies. If a supplier does break away from its parent company, it is often ostracized by other Japanese firms.[6]

Other evidence for a lack of individual rights also exists. Several people argue that Japanese law does not protect the rights of its individuals. According to van Wolferen, Japan allows limited individual rights only to the extent that they further the aims of the state. Most Japanese think of law as an instrument that the Japanese government uses to impose its will. Behind this, van Wolferen argues, is a nonindependent judiciary that is manipulated by government bureaucrats to further the goals of the state.[7]

Status of Women

One group that particularly suffers from a lack of individual rights is Japanese women. Although women in many countries typically have fewer professional opportunities than men, Japanese women have a particularly limited role in their society when compared to women in Western countries. Japanese women are expected to work only until they are married and have children. Women who do work have very little responsibility or authority. Even women with college degrees are paid very little and are often required to serve tea to their male colleagues.

The heavy overtime worked by Japanese men has a strong impact on women. Women rarely see their husbands. Not only do the husbands work long hours, they also tend to socialize a great deal with their coworkers. While the wives stay home (or visit friends in the same situation), Japanese men spend their evenings drinking in a variety of establishments, the most popular being the hostess bar. In these bars, the patrons pay a heavy surcharge on drinks in return for the company of attractive women who listen to their stories, flirt with them, and sing songs with them. It's also common for salarymen to have affairs with these women or to visit the prostitutes who are located in the same parts of town. Such affairs are facilitated by "love hotels," which are found on almost every street in these parts of the city. Sex trips to Thailand and other Southeast Asian countries are also common.[8]

The low esteem in which women are held also affects women in other ways. For example, it is common for women to be fondled on crowded trains where they are often unable to escape; women are expected to endure this humiliation without fighting back since this would cause "embarrassment."[9] Women are also expected to endure bad marriages, which is one reason for the relatively low divorce rate. Most divorces occur either early or late in the marriage. Early divorces are typically due to the husband's late hours socializing with coworkers or his affairs with prostitutes or barmaids. Late divorces often occur after the husband retires and the wife realizes she prefers to spend her time with her friends. Women endure these things because Japanese people are expected to conform to their roles, and these are the roles expected of them.

Treatment of Minorities

Japan expects other groups in its society also to endure unfair treatment without complaining. The most notable of these groups include Koreans, *burakumin*, and foreigners (*gaijin*). A large number of Koreans were brought to Japan as forced labor while Korea was occupied by Japan between 1910 and 1945. About one million Japanese-born Koreans now live in Japan. Although they are treated much better than they were before 1945, they are still excluded from most companies, clubs, or any other organizations that are part of mainstream Japanese life.[10] Despite the fact that most of these Koreans, or even their parents, were born in Japan, few have citizenship.

Burakumin are descendants of butchers, leather workers, grave diggers, and members of other "unclean" professions. In the past, their exclusion was so complete that maps did not even show their villages. As Japan modernized, they lost their special communities and their monopolies over their trades, which has actually increased their economic deprivation. Numbering about three million, they are physically indistinguishable from other Japanese. Official family registers on these people are kept by various organizations, however, to prevent proper Japanese from marrying a burakumin.[11]

Foreigners are also treated differently and often unfairly in Japan although the level of this discrimination depends on their country of origin. Due to their group-oriented mentality, Japanese tend to view people according to the economic might of their home country. Therefore, Americans and other Westerners are typically treated much better than people from South America, other Asian countries, or Africa.

Japanese look askance at all foreigners, however, due to their group orientation. As the largest and most general group in Japan, the Japanese people consider non-Japanese to be outside of this largest group.[12] It typically takes years for a non-Japanese to be allowed into a Japanese group. The attitude of Japanese people toward foreigners can be immediately experienced by visiting or living in those parts of Japan where there are few foreigners (e.g., most places outside of Tokyo, Osaka, or Kyoto). Many Japanese feel free to talk about foreigners in front of them, pointing at them, and in general subjecting them to rudeness that they would not think of inflicting on an ordinary Japanese person. (This is one of the disadvantages of understanding the Japanese language.) Even when foreigners ask Japanese people to stop this type of behavior, they often ignore the request.

Children of mixed marriages are also often victims of discrimination. It's common for other Japanese children to harass and bully these children; the English-language newspapers in Japan are filled with stories about this. The typical responses of school officials provide insight into how the Japanese view individual rights and diversity. These officials almost always place the victims at fault, reasoning that the children of the mixed marriages are different and thus invite harassment.

Japanese also treat those Japanese who have spent significant time abroad in a manner similar to the way they treat foreigners. Akio Morita and Shintaro Ishihara make the point that Japanese with foreign experience are often felt to be tainted; if they have been away for too long, many Japanese feel they may have forgotten how to be Japanese.[13] For this reason, many Japanese make a strong effort to not stay away from Japan for long periods for fear of losing their "Japaneseness."

Discrimination against foreigners and Japanese with lengthy foreign experience obviously has an effect on Japan's relations with foreign countries and its management of foreign subsidiaries. Before moving on to these issues, however, I want to emphasize that there are many Japanese who do not share these views or actions. Other Americans and I have experienced the hospitality of many Japanese. In many cases these Japanese treated us well not because they had to but because they wanted to. Much of this positive behavior, ironically, is also influenced by their culture. These Japanese feel responsible for showing Americans and other foreigners the good side of Japan.

International Relations

Japanese discrimination against foreigners and against Japanese who have substantial foreign experience has a strong, albeit indirect, effect on Japan's international relations. First, it is difficult for Japan to play a role in world affairs when most Japanese have such a low opinion of foreigners and the principles that guide Western countries. Second, the Japanese who have the greatest understanding of foreign countries and Western countries — those who have lived abroad — are not listened to very closely with regard to these issues.

There are, however, more fundamental cultural factors that limit Japan's role in world affairs and its understanding of Western principles. According to van Wolferen, the way in which semiautonomous groups share power in Japan makes it difficult for the Japanese government to promise things or take into account foreign interests. Put another way, the web of power is so closely interwoven that there is no room for foreign interests.[14]

There is much evidence for Japan's maintaining a low profile in world affairs. Japan has never fought for a cause such as democracy, either internally or with an ally. Recent events suggest that Japan has little interest in principles that guide the Western world. I was in Japan during the Salman Rushdie and

Tiananmen Square events, for example. Most Japanese had no opinion on the Salman Rushdie event, despite its implications for freedom of speech. Instead they framed it in terms of a Christian-Muslim issue that they felt was of no concern to them. The Tiananmen Square incident was also seen by the Japanese only in terms of Japanese-Chinese relations rather than in terms of freedom of speech or democracy.[15] The argument was that Japan could not criticize China due to Japan's "special relationship" with China. Although this relationship was never defined, it was clear that the Japanese were more concerned with maintaining trade with China than with upholding any principles of democracy.

Other recent events also show how Japan places more emphasis on context or hierarchy of power than on Western principles. For example, Japan's rising position in the hierarchy of nations is causing many Japanese historians to rewrite the history of World War II. These historians claim that the Rape of Nanking never occurred and that the United States "forced" Japan to attack Pearl Harbor. It is highly probably that as the hierarchy of power continues to shift in Japan's favor, these views will become even more popular.

Japanese Management of Foreign Subsidiaries

The negative aspects of Japanese management and society also affect the performance of Japan's foreign subsidiaries. The views of the Japanese people toward individual rights, women, minorities, and international relations affect how Japanese firms operate in foreign countries because many of these views are far less acceptable outside of Japan. Non-Japanese, particularly Westerners, generally expect more individual rights and better treatment of women and minorities. The problem for Japanese companies is in getting their Japanese employees to act differently outside Japan.

Many writers are reaching similar conclusions in analyses of Japanese-managed U.S. subsidiaries. These studies have found that Japanese managers tend to criticize American workers for their different approaches to work but fail to explain what is expected of them. Many of these managers expect American employees to act similar to Japanese employees and to "understand" what is expected of them.[18]

These studies have also found a large number of language and communication problems. Japanese employees often slip into speaking Japanese in heated discussions, despite the presence of American employees who do not understand Japanese. These language problems often exclude Americans from decision-making processes, much more so than with other foreign-owned companies in the United States.

Studies have also found, however, that Americans are often excluded from decisions for other, more culturally based reasons. In some cases, Americans and Japanese have difficulty understanding each other's different approaches to managing a business. In other cases, decisions are made by Japanese employees while socializing at night when American workers are at home with their families. Finally, Americans employees are often excluded from decisions for purely racial reasons. Apparently, many Japanese don't think it is worth including Americans in important decisions.[19]

Japan's discrimination against Japanese who have significant foreign experience also causes problems for Japanese-owned U.S. subsidiaries. It makes it difficult for Japanese companies to become aware of the problems just described, since these experienced Japanese are not listened to with respect. Moreover, this also affects a Japanese company's views of the U.S. market. David Halberstam, for example, describes how Nissan ignored the opinions of its Japanese employees in the United States for many years.[20] He concludes that Nissan would have become successful in the United States much faster if it had listened more closely to the Japanese who were running the U.S. operations.

The result of these problems is poor performance by many Japanese subsidiaries in the United States and other foreign countries. Kenichi Ohmae claims that 85 percent of Japanese firms in the United States were losing money in 1986.[21] Many Japanese firms apparently have trouble keeping their American executives; a 1989 article reported higher than average turnover rates in Japanese-owned U.S. subsidiaries.[22]

This is not to say that Japanese firms aren't experiencing some success in the United States. New United Motor Manufacturing, Inc. (NUMMI) is a good example of a successfully managed U.S. operation, and there are many more success stories. There are an equal number of bad stories, however. Although some of these stories come from Japanese-run factories,[23] most of the previously mentioned problems are in offices or other white-collar settings. It's likely that blue-collar workers have an easier time than white-collar workers in Japanese-owned firms; blue-collar workers generally receive better treatment in Japanese firms than in U.S. firms. Japanese production workers, for example, have more responsibility than U.S. workers and higher pay as a percentage of CEO pay. Until recently, U.S. firms treated blue-collar workers very poorly, and it's understandable that blue-collar workers might be much happier in a Japanese-run firm than in a U.S. firm.[24] These successes mask fundamental problems for Japanese firms, however. White-collar workers, particularly talented employees, are accustomed to being treated very well in U.S. firms. Such people will not easily accept the way Japanese firms often treat U.S. employees.

Japan's Economic Future

Japan's weaknesses in international relations and in managing foreign subsidiaries will prevent Japan from fully dominating the world. As long as the United States, Europe, and other markets continue to demand and in some cases to increase required domestic content in Japanese imports, Japanese firms will be forced to locate more manufacturing plants and design centers in foreign countries. Due to the inability of Japanese firms to effectively manage these foreign subsidiaries, they will on the average be far less competitive than the manufacturing plants and design centers located in Japan. Ironically, the corporate culture behind Japan's success may eventually become the reason for its slowdown, if not eventual decline. As the world becomes more integrated, international management will become even more important and Japan's weakness in this area may adversely affect its competitiveness. The U.S. government should take these issues into account as it makes decisions concerning Japanese operations in the United States. It may be wise to require even more domestic content in Japanese imports.

Japan's weakness in international management will probably not prevent Japan from increasing its shares of high-technology markets in the next 10 to 20 years, however. Local content laws don't cover most of the high-technology industries at this time, and Japan already dominates or is in the process of dominating most of these industries. Japan's competitiveness continues to grow in the semiconductor, computer, and telecommunications industries. Unless U.S. firms in these industries make substantial changes in the way they are managed, they will most likely continue to lose market share.

I believe that Japan will also eventually dominate the world market for software. Japan's superior software design processes is one reason, but another is that Japan's software industry plays a role very similar to other suppliers in Japan: It has close relationships with its customers. Unlike in the United States, where most software is written by fairly independent software houses, most Japanese software is written either by the producers of electronic hardware (e.g., computers, consumer electronics, manufacturing equipment, and telecommunications equipment) or by suppliers who have very close relations with the producers of hardware. These close relationships probably improve the quality of the final products; moreover, the evidence for Japan's software success already exists in the form of these products. Japan would not be successful in CIM or in the computer, consumer electronic, manufacturing equipment, and telecommunications equipment industries if it could not write good software, since these products contain large amounts of it. As Japan continues to increase its world market share for these products, there will be fewer markets for U.S. software. Eventually, U.S. software will begin to fall technologically behind Japanese software; according to one study it is already happening.[25]

It will be difficult, however, for Japanese firms to become successful in the U.S. market for *packaged* personal computer software. Although their software development processes and close relationships between hardware and software development will give them a competitive advantage, the differences between how personal computers are used in Japan and the U.S. will make it hard for Japanese firms to develop appropriate software for the U.S. market. As we saw in Part Five, Japanese firms have a different concept from their U.S. counterparts as to how computers are used in an office. Japanese firms focus on the total system and have trouble conceiving of a computer that individual employees own and use only themselves. Although I believe that the stronger system orientation of Japanese firms will facilitate their development of computer networks within and between offices, it will also make it difficult for Japanese firms to truly understand the software needs of the U.S. personal computer market.

Japan is also becoming stronger in the aircraft components industry; it appears that several Japanese firms will start producing commercial aircraft very soon. It is possible that Japanese firms will become significant competitors to U.S. aircraft producers by the year 2000, because the design and manufacture of aircraft requires more teamwork than almost any other product today. Aircraft contain a tremendous number of parts (more than 100,000 in the large commercial airliners) and a large variety of different technologies; most aircraft are customized for each buyer. The tremendous amount of teamwork needed to design and produce aircraft will eventually provide Japanese firms with a strong competitive advantage. Japan's only weaknesses in this industry are its relatively small domestic market, its low national defense spending, and until recently, its lack of capability in the technologies important to the industry.

Japan may also become stronger in the chemical, plastics, and pharmaceutical industries. Since less teamwork is needed in these industries than in the discrete parts industries, Japanese companies have not had the advantages in these industries that they have had in discrete parts manufacturing. This has reduced their competitiveness in the nondiscrete parts industries and may continue to affect their competitiveness. However, the growing importance of new technologies and, more importantly, the increasing number of different technologies that affect the nondiscrete parts industries will increase the competitiveness of Japanese firms and enable them to become major players in these industries. The increasing number of technologies involved increases the importance of teamwork, and Japanese firms will then have some competitive advantage. In addition, the financial muscle of Japanese firms will enable them to increase their purchases of U.S. technology and U.S. companies as they have been doing for some time.

There are some industries, however, in which it is unlikely that Japanese firms will become major players in world markets. Industries in which there are no clear technological trends or in which customer demands change somewhat haphazardly present serious problems for Japanese firms. Teamwork and Japan's corporate culture are most useful in those industries in which long-term relationships between laboratories, factories, customers, and suppliers can be beneficial. These relationships are not beneficial in industries that lack clear technological trends, since long-term plans are fairly meaningless in these instances. These types of industries include consumer packaged goods, fashion, cosmetics, entertainment, and designer furniture. These industries depend for their success more on a few creative individuals than on teamwork between large groups of people.

Success in these industries, however, will not be sufficient to maintain the standard of living in the United States. Neither can service industries provide the necessary economic growth. As Cohen and Zysman point out, high-technology service industries depend primarily on high-technology manufacturing industries for their business.[26] Without these manufacturing industries, the service industries will provide very little employment and in the long run probably will not be internationally competitive. The result may be that increased Japanese competition will continue to significantly affect the U.S. standard of living. Instead of having an increasing number of new job openings in high-technology industries, the United States may primarily depend on lower technology service and manufacturing industries to produce new jobs.

These changes are already affecting the U.S. standard of living. One study found that American "baby boomers" between the ages of 35 and 44 have less than 60 percent the net worth, on average, that their parents had when they were the same age. The study predicts that these people will have less than 50 percent of the net worth their parents had when they are between the ages of 45 and 54.[27]

CHAPTER TWENTY-NINE

Becoming Better at Teamwork: Lessons for the United States

To RETAIN THEIR competitive positions in the manufacturing industries, particularly the high-technology manufacturing industries, U.S. firms need to develop a corporate culture that supports teamwork. Teamwork has become more important for success in manufacturing industries over the last 70 years; it is likely to become essential in the future. Products will have more parts, product diversity and automation will increase, a greater variety of technologies will affect individual products, and technology will change more rapidly. Strategies to deal with these changes, such as JIT manufacturing, automation, concurrent engineering, incremental improvements, and shorter design and technology development cycles, will also continue to become more important. Teamwork is an essential part of these strategies.

I am not alone in advocating that U.S. manufacturing firms place a greater emphasis on teamwork. Other observers are also preaching a similar message. Some of their writings were mentioned in Chapter 8; I strongly agree with most of their recommendations. Many U.S. firms are experimenting with programs such as Total Quality, and I applaud their efforts. I don't think these authors or firms go far enough, however.

American management needs to recognize that some aspects of U.S. corporate culture inhibit teamwork. Without radically modifying these approaches, it will be difficult, if not impossible, for U.S. firms to implement the amount of teamwork needed to be competitive in the number of high-technology industries necessary to maintain a high standard of living in the United States. Many Americans think of new manufacturing strategies such as just-in-time or concurrent engineering as revolutions in themselves, since radical changes in corporate culture are needed to implement these new corporate policies. These changes are more radical than most Americans want to believe are needed. Before describing what U.S. management must do, it will be useful to review some of the other prescriptions for America's manufacturing industries and link these proposed changes to a corporate culture change as well.

U.S. Management Problems Flow from the Corporate Culture

Many reasons have been cited for the decline in manufacturing competitiveness in the United States and most of these reasons relate to U.S. corporate culture. Three reasons often cited in the popular press for this decline are U.S. management's emphasis on financial control, its short-term orientation, and its poor relations between labor and management. Although each of these factors does have a deleterious effect on U.S. competitiveness, changing them would also require changes in our corporate culture.

Management in the United States emphasizes financial control because this has been management's primary method of integrating different businesses, functions, machines, and specialists. Rather than developing more subjective measures of performance, U.S. firms have emphasized only those measures that can be tied to the corporation's financial results. Accountants and other managerial specialists have developed sophisticated models to show how increases or decreases in direct labor efficiency and equipment utilization have a direct impact on a corporation's bottom line. More important, these models also imply that more subjective and integrative measures such as quality, technological capability, design decisions, and cycle time do not affect the bottom line. Despite the importance of these measures, the financial models cause employees to ignore them.

Japan's corporate culture provides Japanese firms with other, nonfinancial means of integration. It's not that Japanese firms ignore financial measures of performance, but rather that their corporate culture enables them to be less dependent on these measures and to understand their inherent limitations. Product-oriented forms of organization enable smaller organizational units to be responsible for profitability and costs; they also enable managers who are responsible for these measures to stay close to their products and processes. Therefore, Japanese employees and managers can more easily interpret the financial numbers because they are much closer to the processes, the products, and their associated technology than are U.S. managers. Furthermore, when a process does cross organizational lines, responsibilities are shared and formal communication networks are set up to support these shared responsibilities. Therefore, instead of just managing these organizational units via numbers from afar, the people who manage this integration are directly involved with the products, processes, and associated technology.

Management's short-term orientation is also a commonly given reason for U.S. competitive problems. As we saw at the end of Part Six, the long-term orientation of Japanese management stems from its corporate culture. Japanese firms emphasize long-term relationships with customers and suppliers and incremental improvements in processes and products because their corporate culture makes it easy for them to emphasize long-term strategies.

In a similar manner, U.S. corporate culture encourages and makes it easy for U.S. firms to emphasize short-term relationships with suppliers and customers; corporate culture affirms this as the best type of relationship to have with external organizations. U.S. firms have not emphasized process improvements because their corporate culture affirms that results and output are more important than process or method. They have not emphasized incremental improvements because their corporate culture tells them that large, infrequent innovations made by an elite group of specialists are much more important than incremental improvements made by employees on the bottom rung.

The popular press also cites poor relations between labor and management as a reason for the decline of U.S. competitiveness. This is obviously related to corporate culture, since these problems revolve around how labor and management in the United States define their roles. Traditionally, it has been management's job to think and labor's job to do. As Japan has shown us, however, it is much better for management and labor to share responsibility for thinking and doing and, in many cases, for workers to take a lead role in reducing cycle time and improving quality. In all three cases, U.S. firms need to change their corporate cultures if they want to have better labor-management relations, become more long-term oriented, and place less emphasis on financial management.

Prescriptions for Teamwork

Japanese companies provide the best available model for implementing teamwork in a manufacturing firm. Western firms should not try to become exactly like Japanese firms, however;

Japanese firms and Japanese society have too many problems for Western firms to want to mimic them exactly. The United States and its firms are far ahead of Japan in terms of individual rights, treatment of minorities and women, and management of foreign subsidiaries, and these successes should not be sacrificed in an effort to become more economically competitive.[1] I have little fear that that will happen; my concern is the United States will not change fast or far enough to retain an adequate degree of competitiveness in its manufacturing industries. The United States needs to change its corporate culture. Although it is an incredible challenge, many aspects of Japanese corporate culture can be applied to U.S. firms without sacrificing progress in the respectful treatment of individuals.

Emphasize Process Improvement

First, U.S. firms need to place a greater emphasis on process improvement. Lower inventories, higher quality, shorter delivery times, shorter product development times and faster commercialization of new technologies can be achieved only by improving the processes associated with these goals. A business has many processes, but the three most important processes for manufacturing firms are production, product development, and technology development. Although American firms are making great strides toward reducing cycle time in their production processes, many businesses have not even begun to define their product and technology development processes. Corporate culture prevents many businesses from improving those processes. Most U.S. managers want to believe that these processes depend too much on individual creativity and artistry for them to be defined, much less improved.

Focus Process Improvement Teams on the Product

Second, U.S. firms need to organize themselves more by products than by functions. Processes are easier to improve when individuals are responsible for a large percentage of a process. Since a product-oriented organization tends to have smaller numbers of people responsible for a greater number of activities in a process than does a functional organization, it makes it easier for small groups of people to improve the process. Once a firm starts moving toward a product-oriented organization, it can identify the sections or departments that need to share responsibility for managing and improving the process.

Although many firms are using multifunctional teams to improve processes, these types of teams can complicate an organization more than they focus efforts on process improvement. They often add more dotted lines to an already complicated organizational structure; such types of organization often send conflicting signals to employees. Employees may receive verbal encouragement to participate in multifunctional teams but receive monetary encouragement (e.g., pay raises) from their functional managers only for improving functional performance. A product-oriented organization is a more powerful method of implementing teamwork. It can both institutionalize these multifunctional teams and simplify and flatten the organizational structure.

This is harder to implement than it seems, however. Many U.S. managers would oppose such a change, and it is important to recognize the reasons. Since a product-oriented organization pushes responsibilities for profit and loss further down into an organization, many high-level managers are uneasy about such a change. They feel that they will lose control of many decisions, and this violates their concept of good management. Many U.S. managers also rightly realize that such a change might reduce their power or even eliminate their jobs.

Develop Skills to Support Processes

Third, develop people's skills to support processes. This includes more than just technical skills, however. It requires a complete overhaul of the usual concept of employee skills in the United States. Firms in the United States need to capitalize on the fact that employees have almost 100 percent flexibility over the long run. Firms do not have to accept old job categories or the categories of skills produced by U.S. universities. Firms can develop multifunctional employees, through training and job rotation, whose skills exactly match the firm's processes. This includes white-collar employees such as engineers, marketing analysts, and accountants, in addition to blue-collar workers, for whom these concepts are primarily being considered now.

Corporate culture in the United States discourages firms from changing their concept of employee skills, however. Unions tend to oppose reduced job classifications, pay for skills, and other strategies designed to create more multifunctional workers, because they see that these new strategies will require significant changes in the way they deal with management. Even many managers, particularly in technical areas, are committed to the concept of functional specialization. Too many of them believe that other disciplines are not as good as their own. Engineers, for example, typically think of marketing, sales, purchasing, and drafting as technically beneath them in spite of (or perhaps because of) their key exchanges with these functions.[2] Even within engineering, there is a hierarchy of specialties, with electrical engineering at the top and manufacturing engineering at the bottom. These types of attitudes — and culture — discourage employees from developing an understanding of other activities, even though these activities strongly affect their jobs and how they perform them.

Make Problems and Information Visible

Fourth, U.S. firms need to make problems and performance data visible and they need to make information more accessible. Without visibility, problems are more difficult to solve and employees are often not aware of issues that affect their jobs.

Status boards, status forms, and public files are some of the strategies that can be used to make information more visible and accessible.

We're all familiar with the saying, "information is power." The activities associated with this belief adversely affect manufacturing competitiveness in the United States. American managers need to start viewing information as public rather than private property. As opposed to receiving information on a need-to-know basis, employees should receive more information, and *they* should be allowed to decide whether they need to know it. After all, how can a manager know if employees need some particular information until the employees see it for themselves?

Develop Long-term Relationships with Customers and Suppliers

Fifth, U.S. firms need to develop long-term relationships with their customers and suppliers. Instead of looking at themselves as independent of their suppliers and customers, individual businesses need to recognize that, in the long run, it will not be the best-managed individual businesses that will succeed; it will be the best-managed *collection of businesses within a value chain.* If an individual business is not part of a successful value chain — if it does not have successful customers and suppliers and a means to generate process and product innovations across the entire value chain — it will not succeed no matter how well it manages its own segment. U.S. automobile and steel companies, computer and semiconductor companies, and other electronics industries have been learning this lesson the hard way.

Corporate culture in the United States informs us that businesses should be largely independent of each other and can succeed with loose connections between individual businesses in the value chain. Mainstream economists are largely at fault for this fallacy. Their theories assume there are infinite suppliers, infinite customers, and most erroneously, perfect informa-tion. As we saw in Parts Five and Six, information is the primary medium of the product development and technology develop-ment process and the quality of this information depends largely on the relationship that exists between a business and its cus-tomers and suppliers.

Develop Horizontal Communication Links

Sixth, U.S. firms need to develop communication methods that support close relationships with customers and suppliers and that support internal processes. Formal communication net-works between suppliers, customers, and the necessary internal organizations within a business can provide the structure neces-sary to make joint long-term decisions. In addition, open offices and daily standup meetings facilitate communication not only in small groups, but also between small groups. Like concepts of visible management, an open office and daily standup meetings make information easily available to a large number of people. Forget the typed memos and other time-consuming methods — just bring everyone around the boss's desk and quickly sum-marize what's going on.

Corporate culture in the United States discourages this type of communication. American managers believe that information needs to travel up, across, and down in an organization as op-posed to just horizontally. For many managers, their sole job is to distribute information. They often feel like employees are making an end run when information flows horizontally. Such managers will probably not support a change in corporate cul-ture because it threatens to eliminate their work.

The open office and the daily standup meetings (which re-quire an open office) face the largest cultural hurdles in U.S. firms. Most Americans, including myself, see private offices as perks; more important, we see them as the most efficient arrangement for conducting business, because it is easier to think and get things done in a private office.

The relative merits of open versus private offices depend on one's concept of work, however. If work is looked on as individually based (which some types of work are, such as art, music, or literature), then the private office is the better approach. In a manufacturing firm, however, few people add value as individuals. Almost all employees add value only as part of a team. Therefore, interruptions are a small price to pay for improved teamwork, particularly when these interruptions often prevent someone from wasting time doing the wrong things. Furthermore, people become accustomed to the noise in an open office surprisingly quickly.

Emphasize Team-based Strategies in Business and Engineering Schools

Seventh, universities also have a role in encouraging teamwork. For starters, business and engineering schools should emphasize more team-based strategies such as JIT manufacturing, concurrent engineering, incremental improvements, and short design cycles. In addition, they should provide the students with team-based experience by using group projects to teach these strategies. Not only could universities direct some of these projects toward the problems facing local industries, they also might have these students work with university employees to improve some of the university's white-collar processes. Universities might thus reduce the cost of education, in addition to providing students with valuable experience in process improvement.

Encourage Government Advocacy at All Levels

Eighth, federal, state, and local governments can facilitate teamwork at a macro-level. We've seen some of this recently with changes in laws regarding antitrust regulations and government supports for industry consortiums. The more important thing that government can do, however, is to set a national mood for implementing teamwork. For the United States to become more competitive, Americans must recognize that there is a problem and that there are solutions. Only the president and other national leaders can convey this to the entire nation.

Although the challenges are large, the United States still has a number of strengths and the time to capitalize on them and improve its weaknesses. The United States still leads the world in many technological areas. Its laboratories and universities still have some of the finest minds and capabilities in the world. The United States probably never will regain the international economic domination it had in the first half of the twentieth century, but neither will it become a second-class nation. Its future lies somewhere between these two extremes. The results depend to a large extent on our ability to become more adept at teamwork; to accomplish this, we must change our corporate culture.

Notes

Part One

Chapter 1

1. Data presented by Charles Ferguson in "The Competitive Decline of the U.S. Semiconductor Industry," Written Testimony for the Subcommittee on Technology and the Law, Judiciary Committee, U.S. Senate, February 26, 1987.

2. The original rankings were developed by Regina Kelly in "The Impact of Technological Innovation on Trade Patterns," U.S. Department of Commerce, Bureau of Economic Policy and Research, ER-24, December 1977. These rankings were also discussed by Bruce Scott in "National Strategy for Stronger U.S. Competitiveness," *Harvard Business Review*, March-April 1984. I have changed their rankings slightly so that each industry has about the same market size. I expanded one of these industries (electrical equipment) into electrical components, telecommunications equipment, and other electrical equipment. I also combined industrial and agricultural chemicals into one industry and I eliminated engines and turbines because it is so small.

3. Emily T. Smith, "Japan Pulls Ahead in Another Chip Race," *Business Week*, June 29, 1987, p. 105.

4. U.S. Congress, Office of Technology Assessment, *Making Things Better: Competing in Manufacturing*, March 1990.

5. Michael Porter, *The Competitive Advantage of Nations* (New York: Free Press, 1990), Tables 8-1 and 9-2.

6. See, for example, U.S. Congress, Office of Technology Assessment, *Commercializing High Temperature Superconductivity*, June 1988, p. 33.

7. Sheridan Tatsuno, *Created in Japan: From Imitators to World-Class Innovators* (New York: Harper & Row, 1990).

8. Patent creativity is defined by the Venture Economics and CHI Research firms as the number of times a patent is cited as the basis for subsequent patents. *The Institute*, June 1988, published by the Institute of Electrical and Electronics Engineers.

9. There are numerous such studies. For example see J. A. Adam, "Competing in a Global Economy," *IEEE Spectrum*, April 1990, p. 20, or "Innovation, The Global Race," *Business Week*, Special Issue, June 15, 1990, p. 46.

10. Dori Jones Yang, et al., "How Boeing Does It," *Business Week*, July 9, 1990, p. 46.

11. Clyde V. Prestowitz, Jr., *Trading Places: How We Allowed Japan to Take the Lead* (New York: Basic Books, 1987), p. 240.

12. The import-export ratio was developed by Joseph McKinney and Keith A. Rowley, "Trends in U.S. High Technology Trade: 1965-1982," *Columbia Journal of World Business*, Summer 1985.

13. It also depends on exchange rates, but most producers of the high technology products imported by the U.S. are from countries with an equal or higher per capita income than that of the United States.

14. A similar argument is made by Michael L. Dertouzos, Richard K. Lester, Robert M. Solow, and The MIT Commission

on Industrial Productivity, *Made in America: Regaining the Productive Edge* (Cambridge: MIT Press, 1989).

15. Lois Peters makes a similar argument in her analysis of the Japanese and U.S. chemical industry. "Technical Networks Between U.S. and Japanese Industry," published by the Center for Science and Technology Policy, Rensselaer Polytechnic Institute, March 1987.

16. For example, the MIT Commission on Industrial Productivity describes the strengths of the German chemical industries in *The Working Papers of The MIT Commission on Industrial Productivity*, Volume 1 (Cambridge: MIT Press, 1989).

17. Porter, *The Competitive Advantage of Nations*, Table 8-1.

18. Porter, *The Competitive Advantage of Nations*, Chapter 8.

19. Porter, *The Competitive Advantage of Nations*, Table 9-2.

20. For example, between 1965 and 1985, the global output of the electronics industries grew by 13 percent per annum in real terms. By 1985, it equaled the global ouput of the automobile industry and surpassed that of the steel industry. Reported by Jeffrey Hart and Laura Tyson in "Responding to the Challenge of HDTV," *California Management Review,* Summer 1989.

Chapter 2

1. See William Abernathy, Kim Clark, and Alan Kantrow, "The New Industrial Competition," *Harvard Business Review*, September-October 1981, pp. 68 81.

2. Chalmers Johnson, *MITI and the Japanese Miracle: The Growth of Industrial Policy* (Stanford: Stanford University Press, 1982); Prestowitz, *Trading Places*; Ezra Vogel, *Japan as Number One* (Cambridge: Harvard University Press, 1979); Edwin Reischauer, *The Japanese* (Cambridge: Harvard University Press, 1982); Robert Reich, *The Next American Frontier* (New York: Times Books, 1983).

3. James C. Abegglen and George Stalk, Jr., *Kaisha: The Japanese Corporation* (New York: Basic Books, 1985); Richard Schonberger, *Japanese Manufacturing Techniques: Nine Hidden Lessons in Simplicity* (New York: Free Press, 1982); Yasuhiro Monden, *The Toyota Production System* (Atlanta: Industrial Engineering and Management Press, 1983).

4. William Ouchi, *Theory Z: How American Business Can Meet the Japanese Challenge* (Reading, Mass.: Addison-Wesley, 1981); Richard T. Pascale and Anthony G. Athos, *The Art of Japanese Management: Applications for American Executives* (New York: Warner Books, 1981); Lester Thurow, ed., *The Management Challenge: Japanese Views* (Cambridge: MIT Press, 1985).

5. Johnson, *MITI and the Japanese Miracle*; Michael Porter also mentions this use of the postal savings system to allocate funds to specific industries. Porter, *The Competitive Advantage of Nations*, Chapter 8.

6. Prestowitz, *Trading Places*.

7. In the late 1940s, Japan's standard of living was very similar to countries that were then classified as developing countries.

8. See Porter, *The Competitive Advantage of Nations*, Chapter 8.

9. C. K. Prahalad and G. Hamel, "The Core Competence of the Corporation," *Harvard Business Review*, May-June 1990; Porter, *The Competitive Advantage of Nations*, pp. 406-408.

10. Michael Porter argues that Japan gradually let in foreign competition. Porter, *The Competitive Advantage of Nations*, Chapter 8.

11. See, for example Abegglen and Stalk, *Kaisha*, Chapter 7.

12. Karen Pennar, "Japanese Thrift? Stereotype Suffers a Setback," *Business Week*, August 14, 1989, p. 36.

13. Quoted in "The Myth that the U.S. Is Beaten," *Business Week*, June 15, 1990.

14. Robert Hayes, Steven Wheelwright, and Kim Clark, *Dynamic Manufacturing: Creating the Learning Organization* (New York: Free Press, 1990), p. 12.

15. Thomas Rohlen, "Why Japanese Education Works," *Harvard Business Review,*" September-October 1987.

16. Thomas Rohlen, *Japan's High Schools* (Berkeley and Los Angeles: University of California Press, 1983.)

17. For example, Rohlen argues that Japan is useful as a mirror, not as a model. Rohlen, "Why Japanese Education Works."

18. See, for example, Porter, *The Competitive Advantage of Nations,* Chapter 8.

19. Lawrence P. Grayson, "Technology in Japan: Advancing the Frontiers," *Engineering Education,* April/May 1987.

20. Several studies have found that a large percentage of a product's life cycle costs are determined very early in a product's design cycle. For example, see James Nevins and Daniel Whitney, *Concurrent Design of Products and Processes: A Strategy for the Next Generation in Manufacturing* (New York: McGraw-Hill, 1989), p. 3.

21. Japan External Trade Organization, *Japanese Business Facts and Figures,* Tokyo, 1989.

22. Merry White, *The Japanese Educational Challenge: Commitment to Children* (New York: Free Press, 1987), p. 115.

23. White, *The Japanese Educational Challenge,* p. 130.

24. Duke's book, *The Japanese School: Lessons for Industrial America* (New York: Praeger, 1986), is reviewed by Rohlen in "Why Japanese Education Works."

25. Pascale and Athos, *The Art of Japanese Management.*

26. Prestowitz, *Trading Places,* Chapter 3.

27. Pascale and Athos, *The Art of Japanese Management,* p. 125.

28. Pascale and Athos, *The Art of Japanese Management,* Chapters 3 and 4.

29. Pascale and Athos, *The Art of Japanese Management.*

30. Pascale and Athos, *The Art of Japanese Management.*

31. Ouchi, *Theory Z,* Chapters 3 and 4.

32. Ouchi, *Theory Z,* p. 5.

33. Ouchi, *Theory Z*, Chapters 3 and 4.

34. Schonberger, *Japanese Manufacturing Techniques*.

35. Schonberger, *Japanese Manufacturing Techniques*, Chapter 3.

36. Masaaki Imai, *Kaizen: The Key to Japan's Competitive Success* (New York: McGraw-Hill, 1986).

37. Robert Hayes and Steven Wheelwright, *Restoring Our Competitive Edge: Competing Through Manufacturing* (New York: John Wiley & Sons, 1984), Chapter 13.

38. Edwin Mansfield, "Industrial Innovation in Japan and the United States," *Science* 241 (1988): 1771, Table 4.

39. Ramchandran Jaikumar, "Postindustrial Manufacturing," *Harvard Business Review*, November-December 1986.

40. Porter, *The Competitive Advantage of Nations*, pp. 407-408.

41. Dertouzos, et al., *Made in America*.

42. Kim Clark and Takahiro Fujimoto, *Product Development Performance* (Cambridge: Harvard Business School Press, 1990), Chapter 4; George Stalk, Jr. and Thomas M. Hout, *Competing Against Time* (New York: Free Press, 1990), Chapter 4.

43. See, for example, Nevins and Whitney, *Concurrent Design of Products and Processes*.

44. Stephen S. Cohen and John Zysman *Manufacturing Matters: The Myth of the Post-Industrial Economy* (New York: Basic Books, 1987).

45. Quoted in "The R&D Challenge," *Industry Week*, May 4, 1987.

46. Dertouzos, et al., *Made in America*, p. 104.

47. Prahalad and Hamel, "The Core Competence of the Corporation."

48. Porter, *The Competitive Advantage of Nations*, p. 105.

49. Porter, *The Competitive Advantage of Nations*, pp. 101, 105.

50. Prestowitz, *Trading Places*, Chapter 6.

51. Charles H. Ferguson, "From The People Who Brought You Voodoo Economics: Beyond Entrepreneurialism to U.S. Competitiveness," *Harvard Business Review*, May-June 1988.

52. For example, see Richard Schonberger, *World Class Manufacturing: The Lessons of Simplicity Applied* (New York: Free Press, 1986), Appendix: "Honor Roll: The 5-10-20s."

53. For example, see Michael Cusumano, "Manufacturing Innovation: Lessons from the Japanese Auto Industry," *Sloan Management Review*, Fall 1988.

54. "How to Win the Baldrige Award," *Fortune*, April 23, 1990.

55. Hayes, Wheelwright, and Clark, *Dynamic Manufacturing*, p. 52.

56. Several observers have made this point. The Association for Manufacturing Excellence has published a number of studies in its periodical, *Target*, comparing the implementation of JIT manufacturing in U.S. and Japanese firms.

57. Porter, *The Competitive Advantage of Nations*, p. 19.

58. Schonberger, *Japanese Manufacturing Techniques*, Chapter 7.

59. Richard J. Schonberger, *Building a Chain of Customers: Linking Business Functions to Create the World Class Company* (New York: Free Press, 1990), Chapter 3.

60. Abegglen & Stalk, *Kaisha*, p. 136.

61. See, for example, Edwin Reischauer, *The Japanese*.

62. Prestowitz, *Trading Places*, pp. 156-157.

63. Abegglen and Stalk, *Kaisha*, Figure 3-2.

Part Two

1. Researchers have primarily studied the performance of the production process and the concepts of JIT manufacturing have led many of these researchers to find a correlation between cycle time, quality, and cost. Richard Schonberger was one of the first people to make this connection, in *Japanese Manufacturing Techniques* (1982); many others have noted it since. George Stalk argues that cycle time is equally important to new product design and technology development in "Time — The Next

Source of Competitive Advantage," *Harvard Business Review*, July-August 1988.

Chapter 3

1. Robert Ayres, "Complexity, Reliability, and Design: Manufacturing Implications," *Manufacturing Review*, March 1988.

2. Ayres, "Complexity, Reliability, and Design."

3. Ayres, "Complexity, Reliability, and Design."

4. Ayres, "Complexity, Reliability, and Design."

5. Since automobiles in 1900 had only a few thousand parts, each having three to five distinct surfaces, an automobile probably contained no more than 10,000 distinct surfaces. Yet today's memory chips have more than one million transistors and each of these transistors requires numerous lines, each having a couple of surfaces.

6. Jeffrey Miller and Thomas E. Vollmann, "The Hidden Factory," *Harvard Business Review*, September-October 1985.

7. Booz Allen & Hamilton survey reported in Hirotaka Takeuchi and Ikujiro Nonaka, "The New New Product Development Game," *Harvard Business Review*, January-February 1986.

8. For example, Stalk argues this in "Time — The Next Source of Competitive Advantage."

Chapter 4

1. Porter, *The Competitive Advantage of Nations*, p. 40.

2. Nevins and Whitney, *Concurrent Design of Products and Processes*, p. 34.

3. Hal Mather, "The Case for Skimpy Inventories," *Harvard Business Review*, January-February 1984.

4. Robert Hall, *Attaining Manufacturing Excellence* (New York: Dow-Jones Irwin, 1987), p. 41.

5. For example, Ted Kumpe and Piet T. Bolwijn provide many examples of this in their article, "Manufacturing: The New Case for Vertical Integration," *Harvard Business Review*, March-

April 1988. They claim that there is more value added in building a picture tube than in assembling an entire color television.

6. Miller and Vollmann, "The Hidden Factory."

7. Hayes, Wheelwright, and Clark, *Dynamic Manufacturing,* p. 182.

8. There are a number of good books on this subject. Richard Schonberger was one of the first Americans to describe JIT manufacturing, in *Japanese Manufacturing Techniques* (1982). Since then many other books have been written on the subject. One of the best is Monden, *The Toyota Production System.*

9. Hall, *Attaining Manufacturing Excellence,* p. 147.

10. Hayes, Wheelwright, and Clark, *Dynamic Manufacturing,* p. 184.

11. Schonberger, *World Class Manufacturing,* p. 18.

12. Bruce Henderson, "The Logic of Kanban," *Journal of Business Strategy,* Winter 1986.

13. Hayes, Wheelwright, and Clark, *Dynamic Manufacturing,* p. 257.

14. Schonberger, *World Class Manufacturing,* p. 122.

15. Schonberger, *Building a Chain of Customers,* Chapter 3.

16. Hayes, Wheelwright, and Clark, *Dynamic Manufacturing,* p. 242.

17. Richard Walleigh, "What's Your Excuse for Not Using JIT?" *Harvard Business Review,* March-April 1986.

18. Some people might argue that JIT manufacturing has always been important; after all, didn't Henry Ford first apply JIT manufacturing to the production of automobiles? That may be true, but as Bruce Henderson notes, Ford made only a single product, with no variation. JIT is very easy to apply in such a situation. See Henderson, "The Logic of Kanban."

19. See, for example, Porter, *The Competitive Advantage of Nations,* Figure 8-1.

20. Hall, *Attaining Manufacturing Excellence,* p. 117.

21. Charles Savage, "The Challenge of CIM is 80% Organizational," *CIM Review,* Spring 1988.

22. Quoted in "Management Discovers the Human Side of Automation," *Business Week,* September 29, 1986.

23. Savage, "The Challenge of CIM is 80% Organizational."

Chapter 5

1. For example, an engineer does not design zinc oxide, since its chemical constituents are fixed. In a similar manner, an engineer does not design glass. Rather, laboratory process engineers work with customers to define the needed characteristics of the glass manufacturing process.

2. Porter, *The Competitive Advantage of Nations,* p. 103.

3. According to Hayes, et al., most U.S. firms actually had good teamwork between design engineers and manufacturing engineers before and immediately following World War II. See Hayes, Wheelwright, and Clark, *Dynamic Manufacturing,* p. 182.

4. See, for example, Nevins and Whitney, *Concurrent Design of Products and Processes.*

5. See, for example, Nevins and Whitney, *Concurrent Design of Products and Processes,* pp. 5-7.

6. Nevins and Whitney, *Concurrent Design of Products and Processes,* p. 34.

7. Schonberger, *World Class Manufacturing,* Chapter 8.

8. Hayes, Wheelwright, and Clark, *Dynamic Manufacturing,* p. 331.

9. See for example, Takeuchi and Nonaka, "The New New Product Development Game."

10. Hayes, Wheelwright, and Clark, *Dynamic Manufacturing,* p. 336.

11. For example, McKinsey & Co. found that high-technology products that come to market six months late but on budget will earn 33 percent less profit over five years. In contrast, coming out with a product on time but 50 percent over budget reduces

profits by only 4 percent. Reported in "How Managers Can Succeed Through Speed," *Fortune,* February 13, 1989.

12. Clark and Fujimoto, *Product Development Performance,* Chapter 4.

13. Hayes, Wheelwright, and Clark, *Dynamic Manufacturing,* p. 338.

14. Data presented in "The Software Trap: Automate — or Else," *Business Week,* May 9, 1988.

15. "Software: It's a New Game," *Business Week,* June 4, 1990.

16. For example, see "Creating New Software Was Agonizing Task for Mitch Kapor's Firm," *Wall Street Journal,* June 1990, or Watts Humphrey, "Characterizing the Software Process: A Maturity Framework," *IEEE Software,* March 1988.

17. Watts Humphrey, David Kitson, and Tim Kasse, "The State of Software Engineering Practice: A Preliminary Report," Technical Report, Software Engineering Insititute, Carnegie-Mellon University, Pittsburgh, February, 1989.

18. Michael Cusumano, *Japan's Software Factories* (New York: Oxford University Press, 1991), Chapters 3-7.

Chapter 6

1. Thomas Hughes, *Networks of Power* (Baltimore: Johns Hopkins University Press, 1983).

2. Stuart Leslie, *Boss Kettering: Wizard of General Motors* (New York: Columbia University Press, 1983).

3. Alan Kantrow, "Industrial R&D: Looking Back to Look Ahead," *Harvard Business Review,* July-August 1986.

4. A back-of-the-envelope calculation will prove my point. Westinghouse employed 1200 people in their central research laboratory in 1990. Westinghouse spends less on research and development than most firms (about 2.7 percent of sales) and is about the 30th largest manufacturing firm in the United States in terms of sales. Therefore, it's reasonable to assume that the 10

largest U.S. firms probably employ at least three times as many people in their laboratories as Westinghouse, which means that these laboratories have at least 36,000 employees (1200 × 10 × 3 = 36,000).

5. Roland Schmitt, "Successful Corporate R&D," *Harvard Business Review,* May-June 1985.

6. Roland Schmitt, "Successful corporate R&D."

7. Quoted in "The R&D Challenge," *Industry Week*, May 4, 1987.

8. Quoted in "The R&D Challenge."

9. Dorothy Leonard-Barton and William Kraus, "Implementing New Technology," *Harvard Business Review,* November-December, 1985.

10. Roland Schmitt, "Successful Corporate R&D."

11. Quoted in "The R&D Challenge."

12. Quoted in "The R&D Challenge."

13. Quoted in "The R&D Challenge."

14. Porter, *The Competitive Advantage of Nations,* p. 103.

15. Eric von Hippel, *The Sources of Innovation* (New York: Oxford University Press, 1988), p. 4.

16. Dertouzos, et al., *Made in America,* p. 102.

17. Ted Kumpe and Piet Bolwijn, *Manufacturing: The New Case for Vertical Integration.*

18. Porter, *The Competitive Advantage of Nations,* Chapter 3.

19. Hayes and Wheelwright, *Restoring Our Competitive Edge,* p. 262.

20. Yasuji Sekine, "Japan's 21st Century in Science and Technology," (translation) *Journal of the Japanese Institute of Electrical Engineers,* Vol. 109, No. 7, 1989.

21. Paul Kennedy, *The Rise and Fall of the Great Powers* (New York: Random House, 1987).

22. See, for example, D. N. Perkins, *The Mind's Best Work* (Cambridge: Harvard University Press, 1981).

23. James Burke, *Connections* (Boston: Little, Brown, 1978).

24. Burke, *Connections*, Chapter 6.

25. Burke, *Connections*.

26. Personal communication with Richard Hayes, professor of psychology, Carnegie-Mellon University, December 1988.

Chapter 7

1. This is a summary of material presented in Nevins and Whitney, *Concurrent Design of Products and Processes*, pp. 31-34.

2. David Hounshell, *From the American System to Mass Production, 1800-1932* (Baltimore: Johns Hopkins University Press, 1984).

3. See, for example, Clark and Fujimoto, *Product Development Performance*, Figure 9.2.

4. See, for example, "Smart Cars: Fasten Your Seat Belt for a World of Driving Even 007 Would Envy," *Business Week*, June 13, 1988. Other evidence for this can be found in General Motors's acquisition of Hughes Aircraft Company. Apparently a major reason for GM's decision was the increasing importance of aircraft technologies to automobiles.

5. Personal communication with several members of Mitsubishi's Wire Bonder Equipment Group.

6. Charles Ferguson makes a similar point in "From the People Who Brought You Voodoo Economics: Beyond Entrepreneurialism to U.S. Competitiveness."

7. See, for example, Charles Ferguson, "The Competitive Decline of the U.S. Semiconductor Industry," Written testimony for the Subcommittee on Technology and the Law, Judiciary Committee, U.S. Senate, February 26, 1987.

8. For example, James C. Vesely, head of the Xerox design center, claims that "[chip users] have to understand the silicon right from the beginning." Quoted in "Now Chipmakers are Playing in 'Let's Make a Deal'," *Business Week*, October 6, 1986.

9. See, for example, Sheridan Tatsuno, *Created in Japan*, Chapters 10 and 11.

10. The MIT Commission on Industrial Productivity makes a similar argument, referring to it as the integration of the information industries. *The Working Papers of The MIT Commission on Industrial Productivity*, Volume 2 (Cambridge: MIT Press, 1989).

11. See, for example, George Gilder, "The Revitalization of Everything: The Law of the Microcosm," *Harvard Business Review*, March-April 1988.

Chapter 8

1. Many of Frederick Taylor's views have probably been carried to extremes that even he would not approve of. However, most people seem to agree that Taylor's ideas were the source of much of what is typically called modern management. A good summary of Taylor's teachings and its misrepresentations can be found in Hayes, Wheelwright, and Clark, *Dynamic Manufacturing*, Chapter 2.

2. Thomas Peters and Robert Waterman, *In Search of Excellence* (New York: Warner Books 1984), Chapter 2.

3. Hayes, Wheelwright, and Clark, *Dynamic Manufacturing*, p. 100.

4. Schonberger, *Building a Chain of Customers*, p. 8.

5. Wickham Skinner, *Manufacturing: The Formidable Competitive Weapon* (New York: John Wiley & Sons, 1985), p. 79.

6. Gerald Susman and James Dean, "Organizing for Manufacturable Design," *Harvard Business Review*, January-February, 1989

7. Quoted in Otis Port, "Back to Basics," *Business Week*, June 1989 ("Innovation in America" special issue).

8. Hayes, Wheelwright, and Clark, *Dynamic Manufacturing*, p. 101.

9. Imai, *Kaizen*, p. 51.

10. Hall, *Attaining Manufacturing Excellence*, p. 151.

11. Most of the authors quoted in this section advocate the use of statistical process control to make more problems visible.

12. Schonberger, *World Class Manufacturing*, p. 222.

13. See, for example, Hayes, Wheelwright, and Clark, *Dynamic Manufacturing*, p. 260.

14. Hayes, Wheelwright, and Clark, *Dynamic Manufacturing*, p. 331.

15. William L. Shanklin and John Ryans, Jr., "Organizing for High-Tech Marketing," *Harvard Business Review*, November-December 1984.

16. Margaret Graham, "R&D: Lessons from America's Great Experiment," *American Heritage*, May 1987.

17. For example, see Robert T. Keller, "Predictors of the Performance of Product Groups in R&D Organizations," *Academy of Management Journal*, December 1986, p. 715.

18. Hayes, Wheelwright, and Clark, *Dynamic Manufacturing*, Chapter 4.

19. Hayes, Wheelwright, and Clark, *Dynamic Manufacturing*, p. 316.

20. Imai, *Kaizen*, p. 167.

21. Wickham Skinner, "The Productivity Paradox," *Harvard Business Review*, July-August 1986; Hall, "Attaining Manufacturing Excellence," p. 160.

22. Donald Heany and William Vinson, "A Fresh Look at New Product Development," *Journal of Business Strategy*, Fall 1984.

23. Hayes, Wheelwright, and Clark, *Dynamic Manufacturing*, p. 316.

24. Hayes, Wheelwright, and Clark discuss some of the accounting problems in *Dynamic Manufacturing*, Chapter 5.

25. Hayes, Wheelwright, and Clark discuss some of the accounting problems involved with multifunctional projects in *Dynamic Manufacturing*, Chapter 5.

26. See, for example, Keller, "Predictors of the Performance of Product Groups in R&D Organizations," p. 715.

27. Dertouzos, et al., *Made in America*, p. 98.

28. Schonberger, *World Class Manufacturing*, p. 53.

29. Hayes, Wheelwright, and Clark, *Dynamic Manufacturing*, Chapter 12.

30. Takeuchi and Nonaka, "The New New Product Development Game."

31. See "How to Build and Operate a Product-Design Team," *Industry Week*, April 16, 1990.

32. Data presented in the *IEEE Spectrum*, June 1984.

Part Three

1. Examples of visible management were found in post offices and other governmental organizations.

Chapter 9

1. I use the terms departments, sections, and groups since a specific business within a Japanese firm is typically organized into departments, sections, and groups.

2. In fact, in comparing Mitsubishi with a similar U.S. firm, I found that Mitsubishi has more product-oriented departments than the U.S. firm. For example, the power systems business at one U.S. firm is organized by function at the division level, whereas Mitsubishi's power systems business is organized by product type at the department and sometimes the section level.

3. See, for example, Schonberger, *Japanese Manufacturing Techniques*, p. ix; Imai, *Kaizen*, p. 168; and Dertouzos, et al., *Made in America*, Chapter 6.

4. Ouchi, *Theory Z*, Chapter 1.

5. Ouchi, *Theory Z*, p. 66.

6. Hiroshi Takeuchi makes a similar point that employees of Japanese firms develop their own procedures, whereas employees of U.S. firms are typically given the procedures. See "Motivation and Productivity," in Lester C. Thurow, ed., *The Management Challenge* (Cambridge: MIT Press, 1985).

7. Imai, *Kaizen*, p. 16.

8. Dertouzos, et al., *Made in America*, p. 72.

9. Edwin Mansfield, "Industrial R&D in Japan and the United States: A Comparative Study," *American Economic Review* 78 (1988): 223.

10. Abegglen and Stalk, *Kaisha*, pp. 78-79.

11. Schonberger, *Japanese Manufacturing Techniques*, Chapter 5.

12. Schonberger, *Japanese Manufacturing Techniques*.

13. Miller and Vollmann, "The Hidden Factory."

14. Personal communication with Kazuhiro Kawabata, November 30, 1989.

15. Toshihiro Nishiguchi, "Competing Systems of Automotive Components Supply: An Examination of the Japanese 'Clustered Control' Model and the 'Alps' Structure," Massachusetts Institute of Technology, *International Motor Vehicles Program Working Paper*, May 1987, p. 15.

16. Personal communication with Mr. Banjo, Department Manager, December 5, 1989.

17. Personal communication with Hiroshi Honda, group leader of the wire bonder equipment group, July 10, 1989.

18. Personal communication with Mr. Kawabata, October 24, 1989.

19. Personal communication with Mr. Banjo, December 6, 1989.

20. Personal communication with Mr. Hirayama, December 8, 1989.

21. Personal communication with Mr. Nakamura, December 6, 1989.

22. Personal communication with Mr. Nakamura, December 6, 1989.

Chapter 10

1. These equipment status display techniques were first noted by Richard Schonberger in *Japanese Manufacturing Techniques*, Chapter 3.

2. Schonberger also discusses status boards as a form of visual control; see *Building a Chain of Customers*, pp. 42-43.

Chapter 11

1. Clark and Fujimoto, *Product Development Performance,* Chapter 9.

2. *"Kaisha no chorei de yarukoto"* ("What the Morning Meetings Are All About"), *Shukan Yomiuri,* June 3, 1984, p. 127, quoted in Karel van Wolferen, *The Enigma of Japanese Power* (New York: Knopf, 1990), p. 168.

3. There are intriguing parallels between the organization and procedures of Japanese firms and those used in Japanese primary and secondary schools. Merry White observes that a typical Japanese classroom is chaotic; students shout out ideas and possible answers, moving spontaneously from their desks into huddled groups and exclaiming excitedly over a solution. (*The Japanese Educational Challenge*, p. 114.)

4. For example, see Hiroshi Takeuchi, "Motivation and Productivity," in Lester C. Thurow, ed., *The Management Challenge* (Cambridge: MIT Press, 1985).

5. Ouchi, *Theory Z,* p. 66.

6. Personal communication with Mr. Banjo, December 6, 1989.

Chapter 12

1. Dertouzos, et al., *Made in America,* Chapter 6; Ouchi, *Theory Z,* p. 60; Pascale and Athos, *The Art of Japanese Management.*

2. Personal communication with Mr. Yanagisawa, Manager of Training, November 23, 1989.

3. Personal communication with Mr. Yanagisawa, Manager of Training, June 26, 1989.

4. Personal communication with Mr. Hirayama, December 8, 1989.

5. Personal communication with Mr. Hirayama, December 8, 1989.

6. Personal communication with Mr. Honda, August 2, 1989.

7. Personal communication with Dr. Kishida, October 1989.

8. Personal communication with Dr. Kishida, October 1989.

9. Ouchi, *Theory Z*, pp. 104-105.

10. Kazuo Koike, "Internal Labor Markets: Workers in Large Firms," in Taishiro Shirai, ed., *Contemporary Industrial Relations in Japan* (Madison: University of Wisconsin Press, 1983).

11. Hiroshi Takeuchi, "Motivation and Productivity," in Lester C. Thurow, ed., *The Management Challenge* (Cambridge: MIT Press, 1985).

12. Schonberger, *Japanese Manufacturing Techniques*, p. 195; Abegglen and Stalk, *Kaisha*, p. 136.

13. Leonard H. Lynn, Henry R. Piehler, and W. Paul Zahray, "Engineering Careers in Japan and the United States: Some Early Findings from an Empirical Study." Quoted in U.S. Congress, Office of Technology Assessment, *Making Things Better: Competing in Manufacturing*, March 1990.

Part Four

Chapter 14

1. Bernie Berman, Carroll Casteel, David Dorman, Arthur Gonzales, and Larry Lambert, "An Example of Japanese Total Quality Control," *Brief* 78 (Houston: American Productivity & Quality Center, August 1990).

2. For example, see Schonberger, *Japanese Manufacturing Techniques*.

3. Information from Warren Smith's unpublished notes on the wafer department; received in March 1989. Smith was formerly an engineer with Sygnetics, a U.S. semiconductor maker.

4. Miller and Vollman, "The Hidden Factory."

5. Berman, et al., "An Example of Japanese Total Quality Control," pp. 1-3.

6. Warren Smith's personal notes on the wafer department; received in March 1989.

Chapter 15

1. Personal communication with Mr. Yamada, manager of the assembly department, October 20, 1989.

2. Warren Smith's personal notes on the wafer department; received in March 1989.

Chapter 16

1. Personal communication with Mr. Yamada, a manager in the power device department, November 7, 1989.

Part Five

1. Clark and Fujimoto, *Product Development Performance,* Chapter 4.

2. Stalk and Hout, *Competing Against Time,* Chapter 4.

3. Dertouzos, et al., *Made in America,* Chapter 7.

4. Clark and Fujimoto, *Product Development Performance,* Chapter 9.

5. Dertouzos, et al., *Made in America,* Chapter 5.

6. Clark and Fujimoto, *Product Development Performance,* Chapter 8.

Chapter 17

1. Prestowitz, *Trading Places;* Gene Gregory, *Japanese Electronics Technology: Enterprise and Innovation* (New York: John Wiley & Sons, 1985.)

2. These include the individual equipment, assembly equipment, assembly technology, wafer process equipment, and wafer process technology communication meetings.

3. Personal communication with Mr. Kawabata, September 25, 1989.

4. Personal communication with Koji Matsuda, July 18, 1989.

5. Personal communication with Mr. Hirayama, July 20, 1989.

6. Weekly wire bonder group meeting, October 16, 1989.

7. Personal communication with Mr. Ishihara, June 28, 1989.

8. Personal communication with Mr. Ishihara, June 28, 1989.

9. Personal communication with Mr. Ishihara, June 28, 1989.

Chapter 18

1. Personal communication with Mr. Hirayama, June 8, 1989.

2. Cusumano, *Japan's Software Factories.*

3. Sections 2, 3, and 4 of the system specifications outline the subassemblies and their design conditions.

4. Sections 5, 6, 7, and 8 of the system specifications describe the structure of the hardware and the software, including the number of PCBs, the function of each PCB, and a block diagram of each PCB.

5. Personal communication with Mr. Hirayama, June 8, 1989.

6. Personal communication with Mr. Hiroki, August 24, 1989, and Mr. Yoshitomi, August 2, 1989.

Chapter 19

1. See, for example, Nevins and Whitney, *Concurrent Design of Products and Processes,* pp. 31-34.

2. Presentation by Mr. Hayashi, September 4, 1989.

3. Presentation by Mr. Hayashi, September 4, 1989.

4. Personal communication with Mr. Kawabata, November 14, 1989.

5. Lunch meetings during the month of October 1989.

6. Personal communication with Mr. Kawabata, November 14, 1989.

7. Personal communication with Mr. Kawabata, November 14, 1989.

8. Personal communication with Mr. Kawabata, November 14, 1989.

9. Personal communication with Mr. Kawabata, November 9, 1989.

10. Personal communication with Mr. Nakamura, December 6, 1989.

11. During the week of the meetings with the suppliers, they were discussed each day in the standup meetings.

12. Personal communication with Kazuhiro Kawabata, November 9, 1989.

13. Personal communication with Kazuhiro Kawabata, November 30, 1989.

Chapter 20

1. Personal communication with Mr. Nakamura, November 30, 1989.

2. Personal communication with Mr. Honda, November 29, 1989.

3. Personal communication with Mr. Honda, November 29, 1989.

4. Personal communication with Mr. Hirayama, November 27, 1989.

5. They argue that a significant fraction of Japan's manufacturing competitiveness can be attributed to focused factories that only manufacture narrow product lines. They found that overhead costs in truck manufacturing plants increase as a func-

tion of the number of different truck models manufactured in each plant. (Abegglen and Stalk, *Kaisha.*)

6. Nippon Denso can make 158 different styles of gauge (6 parts) with only 17 different parts. A previous design required 48 different parts to make these different styles. M. M. Andreasen, S. Kahler, and T. Lund, *Design for Assembly* (London: IFS, 1983).

7. Personal communication with Mr. Matsuda, November 1, 1989.

8. Personal communication with Mr. Hiroki, August 23, 1989.

9. Personal communication with Mr. Hiroki, August 23, 1989.

10. Personal communication with Mr. Matsuda, November 2, 1989.

11. Personal communication with Mr. Hiroki, November 4, 1989.

12. Personal communication with Mr. Matsuda, November 2, 1989.

13. Personal communication with Mr. Matsuda, November 29, 1989.

14. Personal communication with Mr. Matsuda, November 29, 1989.

15. Personal communication with Mr. Hirayama, November 24 and 27, 1989.

16. Personal communication with Mr. Hirayama, November 27, 1989.

17. Personal communication with Mr. Hirayama, July 20, 1989.

18. Personal communication with Mr. Hirayama, July 20, 1989.

19. Personal communication with Mr. Hirayama, July 20, 1989.

20. Personal communication with Mr. Hirayama, September 9, 1989.

21. Personal communication with Mr. Nakamura, November 29, 1989.

22. Personal communication with Mr. Nakamura, November 29, 1989.

23. Personal communication with Mr. Hirayama, September 9, 1989.

24. Personal communication with Mr. Hirayama, September 26, 1989.

25. Personal communication with Mr. Kawabata, November 9, 1989.

26. Personal communication with Mr. Honda, November 29, 1989.

27. Personal communication with Mr. Honda, November 29, 1989.

28. Personal communication with Mr. Honda, November 29, 1989.

Chapter 21

1. Mr. Hirayama, internal software report, August 1989.

2. Personal communication with Mr. Banjo, December 1989.

3. Personal communication with Mr. Banjo, December 1989.

4. Mr. Hirayama, internal report, August 1989.

5. Personal communication with Mr. Matsuda, October 14, 1989.

6. Personal communication with Mr. Yoshitomi, August 24, 1989.

7. Personal communication with Mr. Matsuda, October 24, 1989.

8. Personal communication with Mr. Matsuda, August 30, 1989, and Mr. Hirayama, September 7, 1989.

9. Mr. Hirayama, internal software report, August 1989.

10. Mr. Hirayama, internal software report, August 1989.

11. Personal communication with Mr. Matsuda, November 17, 1989.

12. Personal communication with Mr. Matsuda, August 30, 1989.

13. Mr. Hirayama, internal report, August 1989.

14. Personal communication with Mr. Hirayama, October 25, 1989.

15. Personal communication with Mr. Honda, November 29, 1989.

16. Personal communication with Mr. Hirayama, October 25, 1989.

17. Personal communication with Mr. Hirayama, October 31 and November 7, 1989.

18. Personal communication with Mr. Hirayama, October 31 and November 7, 1989.

19. Personal communication with Mr. Hirayama, October 25, 1989.

20. Personal communication with Mr. Hirayama, October 25 and November 7, 1989.

21. Personal communication with Mr. Hirayama, November 7, 1989.

22. Personal communication with Mr. Hirayama, November 7, 1989.

23. Personal communication with Messrs. Matsuda, Banjo, Honda, Yanagisawa, and Nakamura, December 6, 1989, and Mr. Hirayama, December 8, 1989.

24. Personal communication with Messrs. Matsuda, Banjo, Honda, Yanagisawa, and Nakamura, December 6, 1989, and Mr. Hirayama, December 8, 1989.

25. Personal communication with Messrs. Matsuda, Banjo, Honda, Yanagisawa, and Nakamura, December 6, 1989, and Mr. Hirayama, December 8, 1989.

26. Personal communication with Messrs. Matsuda, Banjo, Honda, Yanagisawa, and Nakamura, December 6, 1989, and Mr. Hirayama, December 8, 1989.

27. Personal communication with Mr. Banjo, December 7, 1989.

Part Six

1. Prahalad and Hamel, "The Core Competence of the Corporation."

2. Dertouzos, et al., *Made in America,* Study A, p. 183.

Chapter 22

1. See, for example, Hajime Eto, "Research and Development in Japan," in Yasuhiro Monden, et al., eds., *Innovations in Management: The Japanese Corporation,* Norcross, Ga.: Industrial Engineering and Management Press, 1985.

2. Personal communication with Mr. Honda, September 5, 1989.

3. Personal communication with Mr. Yamada, November 7, 1989.

4. Personal communication with Mr. Yamada, November 7, 1989.

5. Personal communication with Mr. Yanagisawa, December 6, 1989.

6. For example, as discussed in Part Five, Mitsubishi's semiconductor equipment department and several of Mitsubishi's semiconductor chip factories have a formal communication network for the purpose of developing appropriate semiconductor equipment for Mitsubishi's semiconductor factories.

7. Written communication with Mr. Watanabe, July 1989.

8. Written communication with Mr. Kawato, October 1989.

9. Robert Cutler, "A Comparison of Japanese and U.S. High-Technology Transfer Practices," *IEEE Transactions in Engineering Management,* February 1989.

10. Personal communication with Mr. Honda, September 5, 1989.

11. Due to lifetime employment, the number of employees primarily depends on the number of new hires and the number of rotated employees.

12. Personal communication with Mr. Yamada, section manager in the power device department, December 5, 1989.

13. Personal communication with Dr. Koji Kishida, November 20, 1989.

14. Personal communication with Dr. Koji Kishida, November 20, 1989.

15. This generalization is based on my experience in a research laboratory. Obviously, laboratories will differ depending on the management philosophy of the parent firm. However, I think this generalization is at least appropriate for those laboratories that are part of diversified firms that cannot or do not define their future in terms of specific technologies. Without a technological plan for the entire firm, laboratories are discouraged from integrating individual section and department strategies, and that makes these sections and departments autonomous. Each section and department applies and competes with other sections for government, division, and corporate funds.

16. Personal communication with Dr. Koji Kishida, November 20, 1989.

17. Personal communication with Dr. Koji Kishida, November 20, 1989.

18. Edward M. Miller and K. C. Tran, "Factors for Successful Technology Transfer in Mitsubishi's Product Development Process," Westinghouse Technical Report, March 1, 1989.

Chapter 23

1. Howell A. Hammond, "Innovation in the Japanese Company," Conference Board presentation, March 6, 1990.

2. Berman, et al., "An Example of Japanese Total Quality Control."

3. For example, see Dertouzos, et al., *Made in America.*

4. Gregory, *Japanese Electronics Technology*, pp. 68, 90, 251.

5. Charles Ferguson, "From the People Who Brought You Voodoo Economics."

6. Sheridan Tatsuno, *Created in Japan*, Chapters 5 and 6.

7. Richard Rosenbloom and Michael Cusumano, "Technological Pioneering and Competitive Advantage: The Birth of the VCR Industry," *California Management Review*, Summer 1987.

8. See, for example, Hayes, Wheelwright, and Clark, *Dynamic Manufacturing*, Chapter 10.

9. Personal communication with Mr. Matsuda, November 14, 1989.

10. Personal communication with Mr. Ikegami, manager of the deposition and etching equipment section, November 29, 1989.

11. Presentation made at the Fukuoka Works, November 9, 1989.

12. Personal communication with Koji Matsuda, November 7, 1989 and file containing technology maps.

13. Personal communication with Mr. Banjo, December 7, 1989.

14. Presentation by Mr. Ikegami, November 9, 1989.

Chapter 24

1. Personal communication with Mr. Hokamura and Mr. Hori, September 20, 1989.

2. Mitsubishi Technical Report, "High Speed Wire Bonder Development Plan," Manufacturing Development Laboratory, Automation Technology Department.

3. Personal communication with Mr. Kawabata, May 15, 1989.

4. Personal communication with Mr. Hori, September 20, 1989.

5. Written communication with Mr. Kaihara, manufacturing development laboratory, October 1989.

6. Personal communicaton with Mr. Kawabata, October 20, 1989.

7. Personal communication with Mr. Nakasu, November 17, 1989.

8. Personal communication with Mr. Hiroki, November 30, 1989.

9. For example, Hayes, Wheelwright, and Clark make this point in *Dynamic Manufacturing*, p. 324.

10. Personal communication with Mr. Kimura, November 24, 1989.

11. Personnel communication with Mr. Kawato, May 18, 1989.

12. Personal communication with Mr. Miyazaki, September 28, 1989.

13. Personal communication with Mr. Miyazaki, September 28, 1989.

14. Personal communication with Naoki Miyazaki, November 16, 1989.

Chapter 25

1. The number of wire bonders in a packaging system depends on the number of leads in the package. Although the processing time for a die bonder and a mold machine is somewhat independent of package type, for a wire bonder, the greater the number of leads, the more wires that need to be attached to a single package.

2. Personal communication with Mr. Yoshitomi, August 2, 1989.

3. Personal communication with Mr. Hiroki, September 7, 1989.

4. Personal communication with Mr. Yokoyama, September 7, 1989.

Chapter 26

1. Ouchi, *Theory Z*, p. 48.

2. Gregory, *Japanese Electronics Technology*, Chapter 14.

3. "Why NEC Has U.S. Companies Shaking in Their Boots," *Business Week*, March 26, 1990.

4. Ken-ichi Imai, Ikujiro Nonaka, and Hirotaka Takeuchi, "Managing the New Product Development Process: How Japanese Companies Learn and Unlearn," in Kim B. Clark, Robert Hayes, and Christopher Lorenz (eds.), *The Uneasy Alliance: Managing the Productivity-Technology Dilemma* (Cambridge: Harvard Business School Press, 1985).

5. Michael Cusumano, *The Japanese Automobile Industry: Technology and Management at Nissan and Toyota* (Cambridge: Harvard University Press, 1985).

6. Dertouzos, et al., *Made in America*, Study E.

7. Michael Porter, *The Competitive Advantage of Nations*.

8. U.S. Congress, Office of Technology Assessment, *Making Things Better: Competing in Manufacturing*, March 1990, p. 139.

9. Dertouzos, et al., *Made in America*, p. 105.

10. Dertouzos, et al., *Made in America*, p. 107.

11. According to an article in *Business Week*, MITI is planning to provide $167 million of funding in a five-year program for these technologies whereas the National Science Foundation and U.S. industry provide only a few million dollars a year in funding. "Japan Pours Big Bucks into Very Little Machines," *Business Week*, August 27, 1990.

12. This term was developed by Michael Porter in *The Competitive Advantage of Nations*, p. 100.

13. Porter, *The Competitive Advantage of Nations*, pp.100-104.

14. Porter, *The Competitive Advantage of Nations*, p. 100.

15. See, for example, "21 seiki e no kihon senryaku," Keizai Kikakucho Sogo Keikaku Kyoku, Tokyo: Toyo Keizai Shinposha, 1988.

16. Porter, *The Competitive Advantage of Nations*, p. 103.

17. von Hippel, *The Sources of Innovation*, Chapter 1.

18. D. H. Whittaker, "NC/CNC Penetration in Japanese Factories," Contractor report to the U.S. Congress, Office of Technology Assessment, May 1989.

19. Maryellen R. Kelley and Harvey Brooks, "The State of Computerized Automation in U.S. Manufacturing," John F. Kennedy School of Government, Harvard University, 1988.

20. Robert Hof and Neil Gross, "Silicon Valley Is Watching Its Worst Nightmare Unfold," *Business Week*, September 4, 1989.

21. IEEE/USAB Committee on Communications and Information Policy, "U.S. Supercomputer Vulnerability," report to the Institute of Electrical and Electronics Engineers, Inc., prepared by the Scientific Supercomputer Subcommittee, Committee on Communications and Information Policy, United States Activities Board (Washington, D.C., August, 1988); "Today the Chips, Tomorrow the Machines," *Business Week*, July 4, 1988.

22. Dertouzos, et al., *Made in America*, Study E.

23. Ferguson, "From the People Who Brought You Voodoo Economics."

24. Hart and Tyson, "Responding to the Challenge of HDTV."

25. Hart and Tyson, "Responding to the Challenge of HDTV."

26. Hart and Tyson, "Responding to the Challenge of HDTV."

27. Porter's categories are: the Materials/Metals; Forest Products; Petroleum/Chemicals; Semiconductors/Computers; Multiple Business; Transportation; Power Generation and Distribution; Office; Telecommunications; Defense; Food/Beverage;

Textiles/Apparel; Housing/Household; Health Care; Personal; Entertainment/Leisure. See *The Competitive Advantage of Nations,* Figures 7-10, 8-1, 9-4.

Chapter 27

1. U.S. Congress, Office of Technology Assessment, *Commercializing High Temperature Superconductors,* Executive Summary, June 1988; Hart and Tyson, *Responding to the Challenge of HDTV.*

2. R. T. Pascale, "Communication and Decision Making Across Cultures — Japanese and American Companies," *Administrative Science Quarterly,* 23: 91-110, 1978.

3. Clyde Prestowitz describes how Japan's history affects the long-term relationships that exist in Japanese firms. (*Trading Places,* Chapter 6.)

Part Seven

Chapter 28

1. van Wolferen, *The Enigma of Japanese Power;* Jared Taylor, *Shadows of the Rising Sun* (Tokyo: Charles Tuttle, 1985).

2. For example, see Gary Katzenstein, *An Outsider's Year in Japan* (New York: Soho Press, 1989).

3. van Wolferen, *The Enigma of Japanese Power,* pp. 6-8.

4. van Wolferen, *The Enigma of Japanese Power,* p. 393.

5. Due to Japan's incredibly dynamic economy, many engineers have been forced to change from companies in slow growth industries (e.g., chemicals) to companies in high growth industries (e.g., electronics).

6. See, for example, Kuniyasu Sakai, "The Feudal World of Japanese Manufacturing," *Harvard Business Review,* November-December 1990.

7. van Wolferen, *The Enigma of Japanese Power,* pp. 213-226.

8. Taylor, *Shadows of the Rising Sun,* Chapter 7.

9. Taylor also makes this point in *Shadows of the Rising Sun,* Chapter 7.

10. The English newspapers in Japan often have articles describing the unsuccessful attempts by Koreans to obtain membership in various organizations through the Japanese legal system.

11. van Wolferen, *The Enigma of Japanese Power,* p. 74.

12. Many people have noted that the Chinese characters that form the Japanese word for foreigner, *gaijin,* literally mean outside person.

13. Akio Morita and Shintaro Ishihara, *The Japan that Can Say No* (Tokyo: Konbusha, 1989), p. 23.

14. van Wolferen, *The Enigma of Japanese Power,* Chapter 16.

15. See, for example, Toru Yano, "Japan, China Deeply Involved," (editorial) *The Japan Times,* July 11, 1989.

16. See, for example, Douglas Frantz, "Roles of Working Women, Minorities Pose Challenge," *Los Angeles Times,* July 13, 1988.

17. See, for example, Sam Jameson and Karl Schoenberger, "Japanese Give U.S. Workers a Mixed Review," *Los Angeles Times,* July 11, 1988.

18. See, for example, "Japanese Employers Are Locking Out Their U.S. Managers," *Business Week,* May 7, 1990. It quotes a study by the University of Michigan School of Business Administration. Another example is "Culture Shock at Home: Working for a Foreign Boss," *Business Week,* December 17, 1990.

19. Halberstam, David, *The Reckoning* (New York: William Morrow, 1986).

20. Quoted in Stanley J. Modic, "Myths About Japanese Management," *Industry Week,* October 5, 1987.

21. See, for example,"Help Wanted, Room to Advance — Out the Door," *Business Week,* October 30, 1989.

22. See, for example, John A. Byrne, "At Sanyo's Arkansas Plant the Magic Isn't Working," *Business Week,* July 14, 1986.

23. See, for example, Aaron Bernstein, Dan Cook, and Gregory L. Miles, "The Difference Japanese Management Makes, *Business Week,* July 14, 1986.

24. M. A. Harrison, E. F. Hayes, J. D. Meindl, J. H. Morris, D. P. Siewiorak and R. M. White, *JTEC Panel Report on Advanced Computing in Japan* (Baltimore: Loyola College, 1990).

25. Cohen and Zysman, *Manufacturing Matters.*

26. "Those Aging Boomers," *Business Week,* May 20, 1991, p. 108.

Chapter 29

1. In fact, if Japan were able to improve itself in these areas, Japanese firms would become even more competitive.

2. In a similar manner, marketing, sales, and purchasing personnel often think of engineers as too narrow-minded in their outlook on business.

Select Bibliography

Books

Abegglen, James C., and George Stalk, Jr., *Kaisha: The Japanese Corporation*. New York: Basic Books, 1985.

Andreasen, M. M., S. Kahler, and T. Lund, *Design for Assembly*. London: IFS, 1983.

Burke, James, *Connections*. Boston: Little, Brown, 1978.

Clark, Kim, and Takahiro Fujimoto, *Product Development Performance*. Cambridge: Harvard Business School Press, 1990.

Cohen, Stephen S., and John Zysman *Manufacturing Matters: The Myth of the Post-Industrial Economy*. New York: Basic Books, 1987.

Cusumano, Michael, *The Japanese Automobile Industry: Technology and Management at Nissan and Toyota*. Cambridge: Harvard University Press, 1985.

—, *Japan's Software Factories* New York: Oxford University Press, 1991.

437

Dertouzos, Michael L., Richard K. Lester, Robert M. Solow, and The MIT Commission on Industrial Productivity, *Made in America: Regaining the Productive Edge*. Cambridge: MIT Press, 1989.

Duke, Benjamin, *The Japanese School: Lessons for America*. New York: Praeger, 1986.

Gregory, Gene, *Japanese Electronics Technology: Enterprise and Innovation*. New York: John Wiley & Sons, 1985.

Halberstam, David, *The Reckoning*. New York: William Morrow, 1986.

Hall, Robert, *Attaining Manufacturing Excellence*. New York: Dow-Jones Irwin, 1987.

Hayes, Robert, and Steven Wheelwright, *Restoring Our Competitive Edge: Competing Through Manufacturing*. New York: John Wiley & Sons, 1984.

Hayes, Robert, Steven Wheelwright, and Kim Clark, *Dynamic Manufacturing: Creating the Learning Organization*. New York: Free Press, 1990.

Hounshell, David, *From the American System to Mass Production, 1800-1932*. Baltimore: Johns Hopkins University Press, 1984.

Hughes, Thomas, *Networks of Power*. Baltimore: Johns Hopkins University Press, 1983.

Imai, Masaaki, *Kaizen: The Key to Japan's Competitive Success*. New York: McGraw-Hill, 1986.

Japan External Trade Organization, *Japanese Business Facts and Figures*. Tokyo: JETRO, 1989.

Johnson, Chalmers, *MITI and the Japanese Miracle: The Growth of Industrial Policy*. Stanford: Stanford University Press, 1982.

Katzenstein, Gary, *An Outsider's Year in Japan.* New York: Soho Press, 1989.

Kennedy, Paul, *The Rise and Fall of the Great Powers.* New York: Random House, 1987.

Leslie, Stuart, *Boss Kettering: Wizard of General Motors.* New York: Columbia University Press, 1983.

Monden, Yasuhiro, *The Toyota Production System.* Atlanta: Industrial Engineering and Management Press, 1983.

Morita, Akio and Ishihara, Shintaro, *The Japan that Can Say No.* Tokyo: Konbusha Publishing Ltd., 1989.

Nevins, James, and Daniel Whitney, *Concurrent Design of Products and Processes: A Strategy for the Next Generation in Manufaturing.* New York: McGraw-Hill, 1989.

Ouchi, William, *Theory Z: How American Business Can Meet the Japanese Challenge.* Reading, Mass.: Addison-Wesley, 1981.

Pascale, Richard T., and Anthony G. Athos, *The Art of Japanese Management: Applications for American Executives.* New York: Warner Books, 1981.

Perkins, D. N., *The Mind's Best Work.* Cambridge: Harvard University Press, 1981.

Peters, Thomas, and Robert Waterman, *In Search of Excellence.* Warner Books, New York, 1984.

Porter, Michael, *The Competitive Advantage of Nations.* New York: Free Press, 1990.

Prestowitz, Jr., Clyde V., *Trading Places: How We Allowed Japan to Take the Lead.* New York: Basic Books, 1987.

Reich, Robert, *The Next American Frontier*. New York: Times Books, 1983.

Reischauer, Edwin, *The Japanese*. Cambridge: Harvard University Press, 1982.

Rohlen, Thomas, *Japan's High Schools*. Berkeley and Los Angeles: University of California Press, 1983.

Schonberger, Richard, *Japanese Manufacturing Techniques: Nine Hidden Lessons in Simplicity*. New York: Free Press, 1982.

—, *World Class Manufacturing: The Lessons of Simplicity Applied*. New York: Free Press, 1986.

—, *Building a Chain of Customers: Linking Business Functions to Create the World Class Company*. New York: Free Press, 1990.

Skinner, Wickham, *Manufacturing: The Formidable Competitive Weapon*. New York: John Wiley & Sons, 1985.

Stalk, Jr., George, and Thomas M. Hout, *Competing Against Time*. New York: Free Press, 1990.

Tatsuno, Sheridan, *Created in Japan: From Imitators to World-Class Innovators*. New York: Harper & Row, 1990.

Taylor, Jared, *Shadows of the Rising Sun*. Tokyo: Charles Tuttle, 1985.

Thurow, Lester, ed., *The Management Challenge: Japanese Views*. Cambridge: MIT Press, 1985.

Vogel, Ezra, *Japan as Number One*. Cambridge: Harvard University Press, 1979.

White, Merry, *The Japanese Educational Challenge: Commitment to Children*. New York: Free Press, 1987.

van Wolferen, Karel, *The Enigma of Japanese Power*. New York: Knopf, 1990.

von Hippel, Eric, *The Sources of Innovation*. New York: Oxford University Press, 1988.

Periodical Articles

Abernathy, William, Kim Clark, and Alan Kantrow, "The New Industrial Competition," *Harvard Business Review,* September-October 1981, pp. 68-81.

Adam, J. A., "Competing in a Global Economy," *IEEE Spectrum,* April 1990, p. 20.

Ayres, Robert, "Complexity, Reliability, and Design: Manufacturing Implications," *Manufacturing Review,* March 1988.

Bernstein, Aaron, Dan Cook, and Gregory L. Miles, "The Difference Japanese Management Makes," *Business Week,* July 14, 1986.

Byrne, John A., "At Sanyo's Arkansas Plant the Magic Isn't Working," *Business Week,* July 14, 1986.

"Creating New Software Was Agonizing Task for Mitch Kapor's Firm," *Wall Street Journal,* June 1990.

"Culture Shock at Home: Working for a Foreign Boss," *Business Week,* December 17, 1990.

Ferguson, Charles H., "From the People Who Brought You Voodoo Economics: Beyond Entrepreneurialism to U.S. Competitiveness," *Harvard Business Review,* May-June 1988.

Frantz, Douglas, "Roles of Working Women, Minorities Pose Challenge," *Los Angeles Times,* July 13, 1988.

Gilder, George, "The Revitalization of Everything: The Law of the Microcosm," Harvard Business Review, March-April 1988.

Graham, Margaret, "R&D: Lessons from America's Great Experiment," *American Heritage,* May 1987.

Grayson, Lawrence P., "Technology in Japan: Advancing the Frontiers," *Engineering Education,* April/May 1987.

Hart, Jeffrey and Laura Tyson, "Responding to the Challenge of HDTV," *California Management Review,* Summer 1989.

Heany, Donald, and William Vinson, "A Fresh Look at New Product Development," *Journal of Business Strategy, Fall 1984.*

"Help Wanted, Room to Advance — Out the Door," *Business Week,* October 30, 1989.

Henderson, Bruce, "The Logic of Kanban," *Journal of Business Strategy,* Winter 1986.

Hof, Robert, and Neil Gross, "Silicon Valley Is Watching Its Worst Nightmare Unfold," *Business Week,* September 4, 1989.

"How Managers Can Succeed Through Speed," *Fortune,* February 13, 1989.

"How to Build and Operate a Product-Design Team," *Industry Week,* April 16, 1990.

"How to Win the Baldrige Award," *Fortune,* April 23, 1990.

Humphrey, Watts, "Characterizing the Software Process: A Maturity Framework," *IEEE Software,* March 1988.

"Innovation: The Global Race," *Business Week,* Special Issue, June 15, 1990.

"Innovation," *Business Week,* Special Issue, 1989.

Jaikumar, Ramchandran, "Postindustrial Manufacturing," *Harvard Business Review,* November-December, 1986.

Jameson, Sam, and Karl Schoenberger, "Japanese Give U.S. Workers a Mixed Review," *Los Angeles Times,* July 11, 1988.

"Japan Pours Big Bucks into Very Little Machines," *Business Week,* August 27, 1990.

"Japanese Employers Are Locking Out Their U.S. Managers," *Business Week,* May 7, 1990.

Kantrow, Alan, "Industrial R&D: Looking Back to Look Ahead," *Harvard Business Review,* July-August 1986.

Keller, Robert T., "Predictors of the Performance of Product Groups in R&D Organizations," *Academy of Management Journal,* December 1986.

Kumpe, Ted and Piet T. Bolwijn, "Manufacturing: The New Case for Vertical Integration," *Harvard Business Review,* March-April 1988.

Leonard, Frank S. and W. Earl Sasser, "The Incline of Quality," *Harvard Business Review,* September-October 1982.

Leonard-Barton, Dorothy, and William Kraus, "Implementing New Technology," *Harvard Business Review,* November-December 1985.

Mansfield, Edwin, "Industrial Innovation in Japan and the United States," *Science* 241 (1988): 1771.

"Management Discovers the Human Side of Automation," *Business Week,* September 29, 1986.

Mansfield, Edwin, "Industrial R&D in Japan and the United States: A Comparative Study," *American Economic Review* 78 (1988): 223.

Mather, Hal, "The Case for Skimpy Inventories," *Harvard Business Review,* January-February 1984.

McKinney, Joseph and Keith A. Rowley, "Trends in U.S. High Technology Trade:1965-1982," *Columbia Journal of World Business,* Summer 1985.

Miller, Jeffrey and Thomas E. Vollmann, "The Hidden Factory," *Harvard Business Review,* September-October 1985.

Modic, Stanley J., "Myths About Japanese Management," *Industry Week,* October 5, 1987.

"The Myth that the U.S. Is Beaten," *Business Week,* June 15, 1990.

"Now Chipmakers are Playing in 'Let's Make a Deal'," *Business Week,* October 6, 1986.

Pascale, R. T., "Communication and Decision Making Across Cultures — Japanese and American Companies," *Administrative Science Quarterly,* 23 (1978): 91-110.

Pennar, Karen, "Japanese Thrift? Stereotype Suffers a Setback," *Business Week,* August 14, 1989, p. 36.

Prahalad, C. K. and G. Hamel, "The Core Competence of the Corporation," *Harvard Business Review,* May-June 1990.

"The R&D Challenge," *Industry Week,* May 4, 1987.

Rohlen, Thomas, "Why Japanese Education Works," *Harvard Business Review,* September-October 1987.

Rosenbloom, Richard and Michael Cusumano, "Technological Pioneering and Competitive Advantage: The Birth of the VCR Industry," *California Management Review,* Summer 1987.

Sakai, Kuniyasu, "The Feudal World of Japanese Manufacturing," *Harvard Business Review,* November-December 1990.

Savage, Charles, "The Challenge of CIM Is 80% Organizational," *CIM Review,* Spring 1988.

Schmitt, Roland, "Successful Corporate R&D," *Harvard Business Review,* May-June 1985.

Scott, Bruce, "National Strategy for Stronger U.S. Competitiveness," *Harvard Business Review,* March-April 1984.

Sekine, Yasuji, "Japan's 21st Century in Science and Technology," (translation), *Journal of the Japanese Institute of Electrical Engineers,* Vol. 109, No. 7, 1989.

Shanklin, William L., and John Ryans, Jr., "Organizing for High-Tech Marketing," *Harvard Business Review*, November-December 1984.

Skinner, Wickham, "The Productivity Paradox," *Harvard Business Review*, July-August 1986.

"Smart Cars: Fasten Your Seat Belt for a World of Driving Even 007 Would Envy," *Business Week*, June 13, 1988.

Smith, Emily T., "Japan Pulls Ahead in Another Chip Race," *Business Week*, June 29, 1987, p. 105.

"Software: It's a New Game," *Business Week*, June 4, 1990.

"The Software Trap: Automate — or Else," *Business Week*, May 9, 1988.

Stalk, George, "Time — The Next Source of Competitive Advantage," *Harvard Business Review*, July-August 1988.

Susman, Gerald, and James Dean, "Organizing for Manufacturable Design," *Harvard Business Review*, January-February 1989.

Takeuchi, Hirotaka and Ikujiro Nonaka, "The New New Product Development Game," *Harvard Business Review*, January-February 1986.

"Today the Chips, Tomorrow the Machines," *Business Week*, July 4, 1988.

Walleigh, Richard, "What's Your Excuse for Not Using JIT?" *Harvard Business Review*, March-April 1986.

"Why NEC Has U.S. Companies Shaking in Their Boots," *Business Week*, March 26, 1990.

Yano, Toru, "Japan, China Deeply Involved" (editorial), *Japan Times*, July 11, 1989.

Papers, Reports, and Other Articles

Berman, Bernie, Carroll Casteel, David Dorman, Arthur Gonzales, and Larry Lambert, "An Example of Japanese Total Quality Control," *Brief* 78 (Houston: American Productivity & Quality Center, August 1990).

Eto, Hajime, "Research and Development in Japan," in Yasuhiro Monden, et al., eds., *Innovations in Management: The Japanese Corporation* (Norcross, Ga.: Industrial Engineering and Management Press, 1985).

Ferguson, Charles, "The Competitive Decline of the U.S. Semiconductor Industry," Written Testimony for the Subcommittee on Technology and the Law, Judiciary Committee, U.S. Senate, February 26, 1987.

Hammond, Howell A., "Innovation in the Japanese Company," Conference Board presentation, March 6, 1990.

Harrison, M. A., E. F. Hayes, J. D. Meindl, J. H. Morris, D. P. Siewiorak, and R. M. White, *JTEC Panel Report on Advanced Computing in Japan* (Baltimore: Loyola College, 1990).

Humphrey, Watts, David Kitson, and Tim Kasse, "The State of Software Engineering Practice: A Preliminary Report," Technical Report, Software Engineering Insititute, Carnegie-Mellon University, Pittsburgh, February, 1989.

IEEE/USAB Committee on Communications and Information Policy, "U.S. Supercomputer Vulnerability," report to the Institute of Electrical and Electronics Engineers, Inc., prepared by the Scientific Supercomputer Subcommittee, Committee on Communications and Information Policy, United States Activities Board (Washington, D.C., August 1988).

Imai, Ken-ichi, Ikujiro Nonaka, and Hirotaka Takeuchi, "Managing the New Product Development Process: How Japanese Companies Learn and Unlearn," in Kim B. Clark, Robert Hayes, and Christopher Lorenz (eds.), *The Uneasy Alliance: Managing the Productivity-Technology Dilemma* (Cambridge: Harvard Business School Press, 1985).

Kelley, Maryellen R., and Harvey Brooks, "The State of Computerized Automation in U.S. Manufacturing," John F. Kennedy School of Government, Harvard University, 1988.

Kelly, Regina, "The Impact of Technological Innovation on Trade Patterns," U.S. Department of Commerce, Bureau of Economic Policy and Research, ER-24, December 1977.

Koike, Kazuo, "Internal Labor Markets: Workers in Large Firms," in Taishiro Shirai, ed., *Contemporary Industrial Relations in Japan* (Madison: University of Wisconsin Press, 1983).

Lynn, Leonard H., Henry R. Piehler, and W. Paul Zahray, "Engineering Careers in Japan and the United States: Some Early Findings from an Empirical Study." Quoted in U.S. Congress, Office of Technology Assessment, *Making Things Better: Competing in Manufacturing,* March 1990.

Miller, Edward M., and K. C. Tran, "Factors for Successful Technology Transfer in Mitsubishi's Product Development Process," Westinghouse Technical Report, March 1, 1989.

The MIT Commission on Industrial Productivity, *The Working Papers of The MIT Commission on Industrial Productivity,* Volumes 1 & 2 (Cambridge: MIT Press, 1989).

Mitsubishi Technical Report, "High Speed Wire Bonder Development Plan," (Manufacturing Development Laboratory, Automation Technology Department).

Nishiguchi, Toshihiro, "Competing Systems of Automotive Components Supply: An Examination of the Japanese 'Clustered Control' Model and the 'Alps' Structure," Massachusetts Institute of Technology, *International Motor Vehicles Program Working Paper*, May 1987.

Peters, Lois, "Technical Networks Between U.S. and Japanese Industry," published by the Center for Science and Technology Policy, Rensselaer Polytechnic Institute, March 1987.

U.S. Congress, Office of Technology Assessment, *Making Things Better: Competing in Manufacturing*, March 1990.
—-, *Commercializing High Temperature Superconductivity*, Executive Summary, June 1988.

Takeuchi, Hiroshi, "Motivation and Productivity," in Lester C. Thurow, ed., *The Management Challenge* (Cambridge: MIT Press, 1985).

Whittaker, D. H., "NC/CNC Penetration in Japanese Factories," Contractor report to U.S. Congress, Office of Technology Assessment, May 1989.

APPENDIX A

Research Methodology

I began my research at Mitsubishi with several assumptions about what makes Japan successful. Based on books, journal and magazine articles, and my previous experience at Yoko-gawa, I believed that several strategies such as JIT manufacturing, the focused factory, total quality control, concurrent engineering, standardization, short product development times and close relationships with suppliers, laboratories, and customers were important reasons for Japan's economic success. Although it would be interesting to know which of these strategies are most important, I chose to focus on how Mitsubishi implements them. My plan was to learn this information through interviews, documents, and meetings, using triangulation to validate the information.

Although interviews were the primary source of detailed information, they built off the information gathered in meetings and from documents. Table A-1 summarizes these meetings. In my first two months at Mitsubishi, most of my time was spent in training sessions and plant tours. My training was similar to the type of training given to new departmental employees. I learned about each type of equipment and its purpose, design, and technology, with particular emphasis on wire bonding equipment. This training enabled me to meet the key people in the department and the wire bonder equipment group and to learn about some of the key technical aspects of the equipment.

I started attending the weekly wire bonder equipment meetings and daily standup meetings almost as soon as I arrived in Fukuoka. I probably spent more than 150 hours in these meetings during 1989. In these meetings, I learned about the wire bonder equipment group's and the department's general activities. Equipment development projects, their schedules, equipment problems, customers, suppliers, and laboratories were discussed in these meetings. I took detailed notes at almost every meeting and taped about half of these meetings. Documents were passed out in many of these meetings. Equipment and technology schedules and schedules for individual engineers were passed out during weekly equipment meetings. Department and section strategies, technology plans and reports, quality functional deployment (QFD) matrices, equipment performance reports (yield and downtime), and improvement plans were distributed at various times in 1989. At my request, I was also shown the strategic plans for the department's customers (semiconductor factories).

The public files were also a great source of information; engineers frequently answered my questions by showing me documents from the files. For example, I translated the design methodology document and the documentation for one of the wire bonder's major subassemblies. Both of these documents are stored in the public files and were shown to me as the result of questions about the department's design methodology.

The number of interviews and the topics of these interviews is shown in Table A-1. This table includes only those interviews in which I actually gathered new managerial information. It does not include all of the interviews I conducted, since many of these interviews were used to gather technical information that was needed to ask both the managerial questions and to understand the answers.

The interviews shown in Table A-1 ranged in length from 5 minutes to, in some cases, a few hours. This table does not differentiate between long and short meetings. About 75 of the interviews were longer than 30 minutes and about 25 were longer than one hour; the total time for these interviews was almost 100 hours. If multiple topics were addressed in an interview (which was often the case in long interviews), they are counted as multiple interviews in the table.

The interview topics are organized according to where they appear in the book. Not all of the book sections are shown here, however, since some of the detailed information was gathered in documents, in meetings, or in discussions about the other topics. For example, much of the information in Chapter 3 was gathered in organization charts, documents, and meetings. I also do not include the customers as a specific topic in Table A-1 since information concerning the customers was gathered in meetings, from the customer's strategic plans, and in interviews on individual and group responsibilities, the design methodology, new technology, and equipment installation.

Table A-1. Number of Interviews and Meetings

					Month							
	F	M	A	M	J	J	A	S	O	N	D	Total
Meetings Attended												
Weekly equipment meetings	1	3	4	3	4	4	3	3	4	3	1	33
Daily standup meetings												
Section			1	1	1	4	3	2	3	4	10	4	33
Group			1	17	11	16	16	13	14	10	14	2	114
Department meetings			1					2		1		4
Training sessions	15	12		5			3					35
Plant tours	3		1	2				7	4	5	5	32
Informal meetings		5	7	5	3	3	3	5	15	15	2	63
												243
Interviews												
SED managers					1	1	1	1		2	8	14
Wire bonder group			4	10	22	11	15	15	12	12	3	104
Etching equipment group					3	3	7		4	11	3	31
Development group			2	2	7	15	4	6	6	7	2	51
Lead process group					2	1	1				1	5
Laboratories			2					3		4	2	11
General affairs department					1					2	3	6
Suppliers								3	1	1	1	3
												225
Interview Topics												
Employee training		2			1	2				1		6
Individual and group responsibilities					2	7			1	3	10	23
JIT/Production quality		1							2	2		5
Design methodology		2	3	9	3	3	2	3			3	28
Design for manufacturing						1				2	3	6
Suppliers				1	1	2	1		2	1	1	9
Standardization					5	5	8	2	2	14		36
Design improvement activities			2	4	3	3	2	5		3	3	25
New technology		1	8	11	7	11	13	5	13		3	72
Equipment installation					3	4	2	6	3			18
												228

Collection and Analysis of Communication Data

Tables 11-2 and 11-3 describe communication patterns and informal intereactions within the wire bonder group of the Mitsubishi Fukuoka Works semiconductor equipment department. The data for these tables were collected as described below.

Communication Patterns (Table 11-2)

Twelve engineers from the semiconductor equipment department were studied. These included Mr. Nakamura, the department manager; Mr. Honda, the section manager; Mr. Kawabata, the assistant section manager; and nine engineers from the wire bonder group: Messrs. Ishizuka, Yamamoto, Nakasu, Hiroki, Yokoyama, Yoshitomi, Egashita, Hayashi, and Miyazaki.

I collected data each day for one week in July (July 17-21, 1989) and one week in September (September 4-8, 1989). At 16 thirty-minute intervals (8:45 through 11:45 A.M. and 1:00 through 5 P.M.) each day, I noted the activities and location of each engineer. Data collection forms developed for this procedure included columns for noting whether each engineer was working alone, with one, two, or three or more others, or in a meeting; whether he was working at a desk or drafting table, at a computer, in the factory, or in another location. Space was also provided at each interval to make other notes about each engineer's activities. Since the office was open, it was easy for me, as a regular employee, to stand, glance around, and note other engineers' activities. The large status boards in the room were also useful in determining whether engineers were at the factory or otherwise out of the office.

For each engineer, totals were tallied and proportionalized for percentage of time: in the office or in the factory, in the office working alone or with others, and with others in meetings or with others in informal conversations. Daily totals across engineers were then derived. Weekly totals were derived from the daily averages; the weekly totals were averaged to determine the numbers reported in Table 11-2. Less than 10 percent of the data (9.7 percent across two weeks) was unaccounted for; that is, less than 10 percent of the time was the researcher unable to locate and document an engineer's activities and location.

Informal Interactions (Table 11-3)

Data were collected on the informal interactions of engineers within the office on two separate days, one in July 1989 and one in September 1989. Observed were the three senior members of the group — Messrs. Nakamura, Honda, and Kawabata — and six other engineers in the wire bonder group: Messrs. Ishizuka, Yamamoto, Yokoyama, Hiroki, Yoshitomi, and Miyazaki.

At 96 five-minute intervals per day (8:30 to 11:55 A.M. and 12:45 to 5:10 P.M.), these engineers were observed and their activities recorded. The occurrence and duration of conversations was noted, as well as those who participated. If a worker was on the phone, in a meeting, or out of the office, that was noted as well. However, since the focus here was on informal communication, only time spent in the office was analyzed.

Again, since I was a regular employee and group member, I was easily able to stand and locate individuals and note the beginnings and endings of meetings. However, I did select days in which I did not have pressing work, since the data collection involved constant monitoring.

The following were analyzed for each engineer for each day:

- Number of hours he was in the office
- Total number of conversations in which he took part
- Number of conversations per hour in which he took part
- Total number of separate individuals with whom he interacted
- Number of separate individuals per hour with whom he interacted
- Total number of interactions (this is greater than number of conversations since several different people might take part in a given conversation)
- Number of interactions per hour
- Number of phone conversations
- Percentage of intervals in the office in which the engineer was engaged in informal interactions

These numbers were then totaled for each engineer and averaged for a daily total; the two daily averages were themselves averaged for the numbers reported in Table 11-3.

About the Author

Jeff Funk is an assistant professor of business at the Pennsylvania State University in State College, Pennsylvania, where he teaches and does research on the management of technology and manufacturing strategy. He spent a year and a half in Japan as a participant-observer of Yokogawa Electric and Mitsubishi Electric. He speaks and reads Japanese and has written many papers on Japanese management, product development, JIT manufacturing, and computer-integrated manufacturing.

Dr. Funk also has over seven years of manufacturing and product development experience in two U.S. manufacturing firms: Hughes Aircraft Co. and Westinghouse Electric. He was a consultant on product development and Total Quality at the Westinghouse Productivity & Quality Center. He has received numerous grants and awards, including two grants from the National Science Foundation and an award for best article in the journal *Technical Communication*.

Index

BOOKS AVAILABLE FROM
PRODUCTIVITY PRESS

Productivity Press publishes and distributes materials on continuous improvement in productivity, quality, customer service, and the creative involvement of all employees. Many of our products are direct source materials from Japan that have been translated into English for the first time and are available exclusively from Productivity. Supplemental products and services include newsletters, conferences, seminars, institutes, in-house training and consulting, audio-visual training programs, and industrial study missions. Call 1-800-394-6868 for our free book catalog.

Function Analysis
Systematic Improvement of Quality and Performance
Kaneo Akiyama

Function Analysis is a systematic technique for isolating and analyzing various functions in order to better design and improve products. This book gives you a solid understanding of Function Analysis as a tool for system innovation and improvement; it helps you design your products and systems for improved manufacturability and quality. It describes how function analysis is used in the office as well as on the shop floor.
269 pages, FA-B226, $59.95
Concurrent Engineering

Shortening Lead Times, Raising Quality, and Lowering Costs
John R. Hartley

By simultaneously examining the concerns of design, production, purchasing, finance, and marketing from the very first stages of product planning, concurrent engineering makes doing it right the first time the rule instead of the exception. An introductory handbook, it gives managers 16 clear guidelines for achieving concurrent engineering and abundant case studies of Japanese, U.S., and European company success stories.
330 pages, Order CONC-B226, $55.00

Productivity Press, Inc., Dept. BK, P.O. Box 3007, Cambridge, MA 02140
Telephone: 1-800-394-6868 Fax: 1-617-864-6286

Variety Reduction Program (VRP)
A Production Strategy for Product Diversification

Toshio Suzue and Akira Kohdate

Here's the first book in English on a powerful way to increase manufacturing flexibility without increasing costs. How? By reducing the number of parts within each product type and by simplifying and standardizing parts between models. VRP is an integral feature of advanced manufacturing systems. This book is both an introduction to and a handbook for VRP implementation, featuring over 100 illustrations, for top executives, middle managers, and R&D personnel.
164 pages, Order VRP-B226, $59.95

TO ORDER: Write, phone, or fax Productivity Press, Dept. BK, P.O. Box 3007, Cambridge, MA 02140, phone 1-800-394-6868, fax 1-617-864-6286. Send check, purchase order, or charge to your credit card (American Express, Visa, MasterCard accepted).

U.S. ORDERS: Add $5 shipping for first book, $2 each additional for UPS surface delivery. Add $5 for each AV program containing 1 or 2 two tapes; add $12 for each AV program containing 3 or more tapes. CT residents add 6% and MA residents 5% sales tax. We offer attractive quantity discounts for bulk purchases of individual titles; call for more information.

INTERNATIONAL ORDERS: Write, phone, or fax for quote and indicate shipping method desired. Pre-payment in U.S. dollars must accompany your order (checks must be drawn on U.S. banks). When quote is returned with payment, your 'order will be shipped promptly by the method requested.

NOTE: Prices are subject to change without notice.

Productivity Press, Inc., Dept. BK, P.O. Box 3007, Cambridge, MA 02140
Telephone: 1-800-394-6868 Fax: 1-617-864-6286